EW 105
SPACE ELECTRONIC WARFARE

For a complete listing of titles in the
Artech House Electronic Warfare Library,
turn to the back of this book.

EW 105
SPACE ELECTRONIC WARFARE

David L. Adamy

ARTECH HOUSE
BOSTON | LONDON
artechhouse.com

Library of Congress Cataloging-in-Publication Data
A catalog record for this book is available from the U.S. Library of Congress.

British Library Cataloguing in Publication Data
A catalogue record for this book is available from the British Library.

Cover design by Chris Stanfa, Andy Meaden (meadencreative.com)

ISBN 13: 978-1-63081-834-0

© 2021
Artech House
685 Canton Street
Norwood, MA 02062

All rights reserved. Printed and bound in the United States of America. No part of this book may be reproduced or utilized in any form or by any means, electronic or mechanical, including photocopying, recording, or by any information storage and retrieval system, without permission in writing from the publisher.
 All terms mentioned in this book that are known to be trademarks or service marks have been appropriately capitalized. Artech House cannot attest to the accuracy of this information. Use of a term in this book should not be regarded as affecting the validity of any trademark or service mark.

10 9 8 7 6 5 4 3

CONTENTS

PREFACE XIII

1
INTRODUCTION 1

1.1	Orbital Relationships	2
1.2	Book Flow	2
1.3	History of Signal Intelligence Satellite Programs	3
1.4	A Note about Orbital Calculations in This Book	4

2
SPHERICAL TRIGONOMETRY 5

2.1	Plane Trigonometry		5
	2.1.1	The Law of Sines	6
	2.1.2	The Law of Cosines for Sides	6
	2.1.3	The Law of Cosines for Angles	7
	2.1.4	Right Plane Triangle	7
2.2	The Spherical Triangle		7
2.3	Trigonometric Relationships in Any Spherical Triangle		8
	2.3.1	Law of Sines for Spherical Triangles	9

	2.3.2	Law of Cosines for Sides	9
	2.3.3	Law of Cosines for Angles	9
	2.3.4	The Right Spherical Triangle	10
	2.3.5	Napier's Rules	10
2.4		Examples of Plane and Spherical Triangles Used in Problems	11

3
ORBIT MECHANICS 15

3.1		Elements of an Elliptical Orbit	15
3.2		Relationship Between the Size of an Orbit and Its Period	18
3.3		SVP	20
3.4		An Earth Surface Location	21
3.5		The Propagation Distance	22
3.6		Earth Traces	23
	3.6.1	Satellite Earth Trace	23
	3.6.2	Polar Orbit Earth Traces	25
	3.6.3	Synchronous Satellite Earth Traces	26
3.7		Location of an EW Threat	28
	3.7.1	Calculating the Look Angles	28
	3.7.2	Calculating the Azimuth to the Threat	29
	3.7.3	Calculating Range and Elevation to the Threat Location	31
3.8		Calculating the Distance to the Horizon	32
	3.8.1	Horizon Distances for Circular Orbits	35

4
RADIO PROPAGATION 37

4.1		Introduction	37
4.2		One-Way Link	37
4.3		Propagation Loss Models	40
	4.3.1	LOS Propagation	42
	4.3.2	Two-Ray Propagation	46
	4.3.3	Minimum Antenna Height for Two-Ray Propagation	48
	4.3.4	A Note about Very Low Antennas	49
	4.3.5	Fresnel Zone	49

4.3.6	Complex Reflection Environment	50
4.3.7	KED	50
4.3.8	Atmospheric Loss	54
4.3.9	Rain and Fog Attenuation	55

5
RADIO PROPAGATION IN SPACE 57

5.1	LOS Loss	57
	5.1.1 Atmospheric Loss	58
5.2	Antenna Misalignment	59
5.3	Polarization Loss	60
5.4	Rain Loss	62

6
SATELLITE LINKS 69

6.1	Link Geometry	69
	6.1.1 Looking Down	72
	6.1.2 Plug in Some Orbit Numbers	74
	6.1.3 Looking Up	75
6.2	Uplinks	76
	6.2.1 Command Links	78
	6.2.2 Intercept Links	79
6.3	Downlinks	80
	6.3.1 Telemetry Link	81
	6.3.2 Data Link	81
	6.3.3 Links to Data Users	82
	6.3.4 Jamming Links	82
6.4	Hostile Links	87

7
LINK VULNERABILITY TO EW 91

7.1	Satellite Vulnerability	91
	7.1.1 Space-Related Link Losses	94

	7.1.2	Intercept	94
	7.1.3	Spoofing	94
	7.1.4	Jamming	94
	7.1.5	Problems Worked in This Chapter	95
7.2	Downlink Intercept		96
	7.2.1	Geocentric Angle from the Satellite to the Intercept Site	96
	7.2.2	Range from the Satellite to the Intercept Site	97
	7.2.3	Is the Satellite Above the Horizon from the Intercept Site?	99
	7.2.4	How Strong Is the Downlink Signal at the Intercept Site	101
	7.2.5	Intercept with Directional Antennas	101
	7.2.6	Angles Relative to the Ground Control Station	103
	7.2.7	Angles Relative to the Hostile Intercept Site	105
	7.2.8	Received Signal at the Intercept Site	109
	7.2.9	What Is the Quality of the Intercepted Downlink Signal?	109
7.3	Intercepting Uplinks		110
7.4	Jamming Downlinks		114
	7.4.1	The Satellite Downlink	115
	7.4.2	The Jammer Link	118
	7.4.3	The J/S Formula	118
7.5	Jamming Satellite Uplinks		119
	7.5.1	The Jamming Link	120
	7.5.2	Jamming Link Loss	122
	7.5.3	Gain of 4-m Jamming Antenna at 5 GHz	122
	7.5.4	Jammer ERP	122
	7.5.5	The Satellite Uplink	123
	7.5.6	Gain of 2-m Uplink Transmitter Antenna	124
	7.5.7	Uplink ERP	124
	7.5.8	Uplink Loss	125
	7.5.9	Gain and Bandwidth of the Uplink Receiving Antenna on the Satellite	125
	7.5.10	Antenna 3-dB Beamwidth	125
	7.5.11	Offset of the Jammer from the Downlink Receiving Antenna	126
	7.5.12	The J/S	127
7.6	Electronic Protection of Satellite Links		128

7.6.1	Attacks on Links		128
7.6.2	Protection Against Jamming		130

8
DURATION AND FREQUENCY OF OBSERVATIONS 133

8.1	Calculating the Distance to the Horizon		134
8.2	Horizon Distances for Circular Orbits		136
8.3	Calculating the Duration of Target Availability from a Satellite		137
	8.3.1	Geocentric Viewing Angle	139
	8.3.2	Time During Which a Satellite Can See a Point on the Earth	141
	8.3.3	The Impact of the Movement of the Earth	142
	8.3.4	Viewing Time Formula	143
8.4	Doppler Shift in Satellite Link		144
	8.4.1	Doppler Shift Formula	144
	8.4.2	Receiving Site Velocity	145
	8.4.3	Satellite Velocity	145
	8.4.4	General Formula for Maximum Doppler Shift	148
	8.4.5	General Formula for the Doppler Shift	148

9
INTERCEPT FROM SPACE 151

9.1	Intercept of Radar Signal from Low-Earth Satellite		151
	9.1.1	Link Loss	152
	9.1.2	LOS Loss	154
	9.1.3	Atmospheric and Rain Loss	154
	9.1.4	Can the Satellite Payload Receive the Signal?	157
	9.1.5	Receiver Sensitivity	159
	9.1.6	Link Margin	159
	9.1.7	Could the Satellite Receive the Signal from Its Horizon?	159
	9.1.8	How Long Will the Satellite See the Signal?	162
9.2	Horizon Plot on the Earth		163
9.3	Intercept of the Earth Surface Target Using a Narrow-Beam Receiving Antenna		166

CONTENTS

9.3.1	Antenna Pointing	166
9.3.2	Intercept Link Equation	169
9.3.3	Link Losses	172
9.3.4	Intercept from the Horizon	176
9.4	Intercept from the Synchronous Satellite	180
9.4.1	With the Satellite on the Horizon	180
9.4.2	With the Satellite Directly Overhead	182

10
JAMMING FROM SPACE 183

10.1	Jamming of a Ground Signal from a Satellite	183
10.2	Jamming from a Satellite	183
10.3	Jamming of a Communications Network	186
10.3.1	The Network	186
10.3.2	Link Equations	186
10.3.3	J/S	190
10.4	Jamming a Microwave Digital Data Link	190
10.5	Jamming of a Ground Radar from Space	194
10.5.1	Radar Jamming from a Satellite	194
10.5.2	The Jammed Radar and Its Target	195
10.5.3	The Jammer	196
10.5.4	The Jamming Equation	197
10.5.5	The Jamming Adequacy	197
10.5.6	Protecting an Asset with Low Radar Cross-Section	197
10.5.7	Duration of Jamming	198

A
FORMULAS FROM SIGNAL INTELLIGENCE AND EW 199

A.1	Intercept Formulas	200
A.1.1	Successful Intercept	200
A.1.2	The Intercept Link Equation	201
A.1.3	Received Signal Quality	201
A.2	Communication Jamming	201
A.2.1	Successful Communications Jamming	202

	A.2.2	Communications J/S	203
	A.2.3	Communications EP	204
A.3	Radar Jamming		205
	A.3.1	Successful Radar Jamming	205
	A.3.2	Self-Protection Jamming	205
	A.3.3	Stand-Off Jamming	206
	A.3.4	Required J/S	207
	A.3.5	Radar EP	207

B
IMPORTANT NUMBERS FOR SPACE EW 211

C
DECIBEL MATH 213

C.1	Decibel Numbers	213
C.2	Conversion to Decibel Form	214
C.3	Absolute Values in Decibel Form	215
C.4	Decibel Forms of Equations	216
C.5	Quick Conversions to Decibel Values	217

BIBLIOGRAPHY	221
ABOUT THE AUTHOR	223
INDEX	225

PREFACE

This book deals with the intersection of two disciplines: electronic warfare (EW) and satellites. It is written in the hope that it will be useful to those who are EW professionals but know little about satellites or who are satellite professionals but know little about EW. It is also designed to be useful to those who are new to both. EW is very real in space. There are many satellites that are vital to both military and civil activities. Many are under current electronic attack, and the rest are vulnerable to attack that may not yet have occurred.

There is an appendix on EW basics and chapters on the basics of spherical trigonometry, orbit mechanics, and radio propagation. This book deals with specific kinds of problems that must be solved. What does it take to intercept a hostile signal from space? What does it take to jam a hostile signal from space? What can an enemy do to keep our satellite from doing its job and what can be done to protect our satellite from that enemy activity?

This book takes a cookbook approach to the description of problems that those in the conduct of space EW may be called upon to solve. Step by step, how can you prepare the desired dish? Each problem is described in terms of the physical nature of the situation. Then an example is presented to show generically the way to determine the required answers. Numbers are plugged in to get numerical answers. The idea is that when you are called upon to solve a real-world problem, you can plug the real-world specifications into the equations to get the required real-world answers. This requires that we wade though some spherical trigonometry, but that is the nature of space EW, and the answers that we get can be a matter of life or death.

1

INTRODUCTION

There is increasing interest in electronic warfare (EW) in space because of the strategic advantages that satellites offer. Because of their elevated positions, they can see a great distance, and they can remain operational for extended periods. Although satellites can be attacked kinetically, it is a great deal of trouble to do so; they are small and far away. This makes them valuable as EW platforms.

A disadvantage of satellites is that they are in general unmanned. This means that communication with a satellite requires an electromagnetic link which is vulnerable to enemy countermeasures. Another disadvantage of a satellite is that it cannot be easily steered to an optimum location. However, we can predict the timing versus satellite location, so we can plan other events with full knowledge of when a satellite will be able to view a part of the Earth's surface.

In general, the higher a satellite is, the longer it can see a specific part of the Earth. Low orbits, barely above the atmosphere, have orbital periods down near 1.5 hours and they only dwell for a few minutes over one potential target, and can see out to about 2,000 km from the sub-vehicle point (the sub-vehicle point is on the Earth right below the satellite). When a satellite is at about 37,000 km altitude, it has an orbital period of 24 hours, so it can hover over one location indefinitely and can see about 45% of the Earth's surface. This is called a synchronous orbit. We will deal with the calculation of these and related values for specific satellite parameters in later chapters.

A major trade-off in the selection of satellite orbit parameters is the range between the satellite and a potential target for intercept or jamming. As stated above, satellites are far away and thus incur very large signal transmission losses.

1.1 ORBITAL RELATIONSHIPS

After some important math related to satellites, we will cover some basic relationships in orbits. We will skip many of the niceties of derivations and minor details in this book, but will go directly to the information required for us to work practical, EW-related problems: primarily intercept and jamming of hostile signals transmitted from the Earth's surface and the vulnerability of satellite links to attack from the Earth's surface. We will divide our attention between hostile radars and hostile communications.

The path of an Earth satellite is affected by the gravitational pull of every other object in space, but most of those objects are far away and thus have only secondary effects. The main elements determining the orbit are the gravitational pull of the Earth and the velocity of the Earth satellite. By ignoring other factors, we have what is called the two-body problem. For our purposes at this time, we assume there is nothing in space except the Earth and one satellite moving around the Earth. Working with the mechanics of the orbit and dealing with angles and distances between satellites and the transmitters and receivers involved in EW operations will require the use of spherical trigonometry, so we will also cover that rather gently, focusing on the equations required to work EW problems. The other important math involves radio propagation. We will cover this both within the atmosphere and in space.

1.2 BOOK FLOW

This book has 10 chapters and 3 appendixes:

- This chapter, Chapter 1 Introduction;
- Chapter 2 Spherical Trigonometry: relationships in spherical triangles;

- Chapter 3 Orbit Mechanics: Earth satellite ephemerides, orbit size versus period, look angles and ranges between satellites and targets, and horizon distances and angles;
- Chapter 4 Radio Propagation: within the atmosphere including formulas for radar and communication transmissions; line of sight, two-ray, and interactions with terrain; atmospheric loss; and rain loss;
- Chapter 5 Radio Propagation in Space: atmospheric and rain loss and Doppler shift between satellite and ground-based station;
- Chapter 6 Satellite Links: command links, data links, telemetry links, payload links, and interference links;
- Chapter 7 Link Vulnerability to EW: intercept and jamming of satellite links, spoofing, and electronic protection of satellite links;
- Chapter 8 Duration and Frequency of Observations: the time that the satellite is over the horizon from targets and the Doppler shift;
- Chapter 9 Intercept from Space: intercept of radar and communication targets from low orbit and synchronous orbit;
- Chapter 10 Jamming from Space: jamming of radar and communications targets from low Earth orbits;
- Appendix A Formulas from Signal Intelligence and EW: communication and radar jamming definitions and formulas;
- Appendix B Useful Data for Space Calculations: constants related to orbits and the Earth, orbit radius, and period tables;
- Appendix C Decibel Math: a review of the way the numbers and formulas are handled in decibel form and the way to convert signal parameters into logarithmic form.

1.3 HISTORY OF SIGNAL INTELLIGENCE SATELLITE PROGRAMS

Because of the nature of the hostilities between Western countries (the United States and Europe) and Eastern countries (primarily the Soviet Union, now Russia, and China), it became very important for each side to know the capabilities and intentions of the other side. The idea was to avoid accidentally getting into a mutually destructive war. After the "Open Skies" initiative of the 1950s (for mutually agreed overflights of aircraft) failed to materialize, military reconnaissance satellites became the most practical way to obtain this information.

In 1994, the National Reconnaissance Office (NRO) published a book about the programs that produced and operated those satellites. It was declassified (with significant redactions) in 2016. It should be noted that this book has lots of information about the programs and the people who managed and participated in them, but it does not have the type of unclassified technical information that is contained in this book.

In 1960, Western countries started launching imaging and signal intelligence satellites. Now there are a significant number of those satellites in operation. The capabilities of these satellites have increased yearly in response to all of the technology advances that have occurred. The purpose of these satellites was (and is) to identify military threats and to determine the capabilities of the weapons and their sensors.

1.4 A NOTE ABOUT ORBITAL CALCULATIONS IN THIS BOOK

There is a lot of trigonometry in this book. This is because the inputs to EW equations are very dependent on the relative positions of satellites and ground transmitters or receivers. Also important to calculations are the relative directions to other "players." When dealing with relationships in spherical and plane triangles in problems presented in all of the chapters, it may seem that a few things are overexplained. This is because I, like many other people, can get lost in the geometry and thus would prefer to overexplain something than to leave out logical steps that seem obvious. This is done without apology.

2

SPHERICAL TRIGONOMETRY

In order to deal with practical problems involving Earth satellites, it is necessary to deal with three-dimensional (3-D) angular relationships. Spherical trigonometry is a necessary tool for determining such values as the elevation and azimuth of look angles to the satellite from the ground or the angles to ground transmitters or receivers from the satellite.

This chapter is not intended as a complete coverage of spherical trigonometry; it only provides the background necessary to deal with the scope of the satellite problems on which we will be working in the later chapters.

We will deal with the basic trigonometry formulas: first plane trigonometry and then spherical trigonometry. Then we will develop formulas for various specific geometric relationships important to EW applications.

2.1 PLANE TRIGONOMETRY

First, let us review some plane trigonometry. Many of the problems covered in this book will use both plane and spherical trigonometric relationships. Plane trigonometry deals with triangles in planes. As shown in Figure 2.1, there are three sides and three angles. The sides have physical length and the sum of the three internal angles of the plane triangle adds to 180°. It is common practice to label the sides of

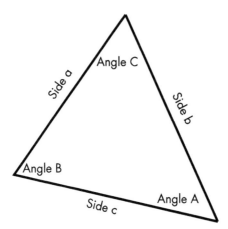

Figure 2.1 The plane triangle is contained within a plane.

a triangle with lowercase letters and the angle opposite each side with the corresponding uppercase letter.

2.1.1 The Law of Sines

In any plane triangle, the relationships between the three sides and three angles are:

$$\frac{a}{\sin A} = \frac{b}{\sin B} = \frac{c}{\sin C}$$

that is, the ratio of the length of a side to the sine of the angle opposite that size is the same as the same ratio for the other three sides.

2.1.2 The Law of Cosines for Sides

The relationship between the sides and angles of any plane triangle is shown by two formulas; this one is called the law of cosines for sides:

$$a^2 = b^2 + c^2 - 2bc \cos A$$

This formula is useful when two sides and an angle are known.

2.1.3 The Law of Cosines for Angles

The other law of cosines is useful when two angles are known. It is called the law of cosines for angles:

$$a = b \cos C + c \cos B$$

These three formulas, along with the spherical trigonometry formulas that follow, will be used in most of the problems in later chapters.

2.1.4 Right Plane Triangle

As shown in Figure 2.2, a right plane triangle is just a plane triangle with one 90° angle.

2.2 THE SPHERICAL TRIANGLE

A spherical triangle is defined in terms of a unit sphere, which is a sphere of radius 1 as shown in Figure 2.3. The origin (center) of this sphere is placed at the center of the Earth in navigation problems, at the center of the antenna in angle-from-boresight problems, and at the center of an aircraft for weapon engagement scenarios. There are an infinite number of applications, but for each, the center of the sphere

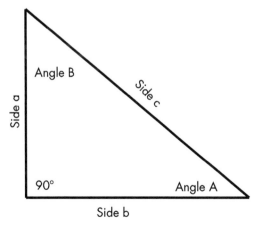

Figure 2.2 The right plane triangle has one angle, which is 90°.

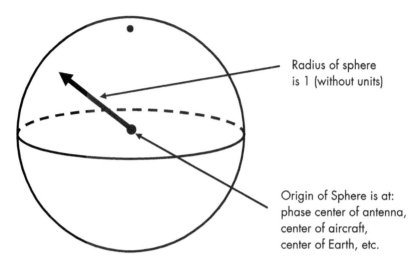

Figure 2.3 Spherical trigonometry is based on relationships in a unit sphere. The origin (center) of the sphere is some point that is relevant to the problem being solved.

is placed where the resulting trigonometric calculations will yield the desired information.

The sides of the spherical triangle must be segments of great circles of the unit sphere; that is, they must be along the intersection of the surface of the sphere with a plane passing through the origin of the sphere. The angles of the triangle are the angles at which these planes intersect. Both the sides and the angles of the spherical triangle are angles. The size of a side is the angle that the two end points of that side make at the origin of the sphere. In normal terminology, the sides are indicated as lowercase letters, and the angles are indicated with the uppercase letter corresponding to the side opposite the angle as shown in Figure 2.4.

It is important to realize that some of the qualities of plane triangles do not apply to spherical triangles. For example, all three of the angles in a spherical triangle could be 90º.

2.3 TRIGONOMETRIC RELATIONSHIPS IN ANY SPHERICAL TRIANGLE

While there are many trigonometric formulas, the three most commonly used in EW applications are the law of sines, the law of cosines for angles, and the law of cosines for sides. In each of these formulas, the lowercase letter is the length of a side and the uppercase letter is

2.3 TRIGONOMETRIC RELATIONSHIPS IN ANY SPHERICAL TRIANGLE

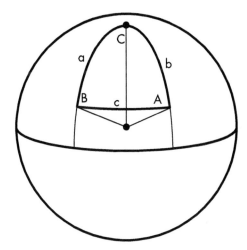

Figure 2.4 A spherical triangle has three sides that are on great circles of a sphere. It has three angles, which are the intersection angles of the planes including those great circles.

the magnitude of the angle opposite that side. We will use degrees for both sides and angles in this book, but be aware that any angular units (e.g., gradients or radians) can be used if more convenient.

These functions are defined as follows.

2.3.1 Law of Sines for Spherical Triangles

$$\frac{\sin a}{\sin A} = \frac{\sin b}{\sin B} = \frac{\sin c}{\sin C}$$

2.3.2 Law of Cosines for Sides

$$\cos a = \cos B \cos C + \sin B \sin C \cos a$$

2.3.3 Law of Cosines for Angles

$$\cos A = -\cos B \cos C + \sin B \sin C \cos a$$

a can be any side of the triangle that you are considering and *A* will be the angle opposite that side. You will note that these three formulas are similar to equivalent formulas for plane triangles.

2.3.4 The Right Spherical Triangle

As shown in Figure 2.5, a right spherical triangle has one 90° angle. This figure is the way that the latitude and longitude of a point on the Earth's surface would be represented in a navigation problem, and many EW applications can be analyzed using similar right spherical triangles.

2.3.5 Napier's Rules

Right spherical triangles allow the use of a set of simplified spherical trigonometric equations generated by Napier's rules. Note that the five-segmented disk in Figure 2.6 includes all of the parts of the right spherical triangle except the 90° angle. Also note that three of the parts are preceded by "co-." This means that the trigonometric function of that part of the triangle must be changed to the cofunction in Napier's rules (i.e., sine becomes cosine).

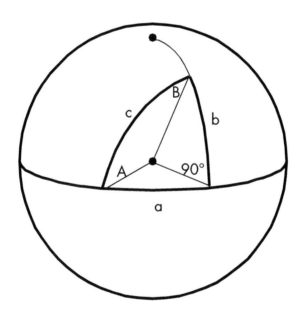

Figure 2.5 A right spherical triangle has one 90° angle.

2.4 EXAMPLES OF PLANE AND SPHERICAL TRIANGLES USED IN PROBLEMS

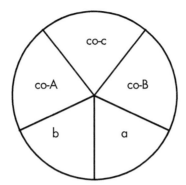

Figure 2.6 Napier's rules for right spherical triangles allow simplified equations in reference to this five-segment circle.

Napier's rules are:

1. The sine of the middle part equals the product of the tangents of the adjacent parts. (Remember the co-s.)
2. The sine of the middle part equals the product of the cosines of the opposite parts. (Remember the co-s.)

A few example formulas generated by Napier's rules are:

$$\sin a = \tan b \, \cotan B$$
$$\cos A = \cotan c \ \tan b$$
$$\cos c = \cos a \ \cos b$$
$$\sin a = \sin A \ \sin c$$

As you will see when they are applied to Earth satellite and other practical EW problems, these formulas greatly simplify the mathamatics involved with spherical manipulations when you can set up the problem to include a right spherical triangle.

2.4 EXAMPLES OF PLANE AND SPHERICAL TRIANGLES USED IN PROBLEMS

Figure 2.7 shows a spherical triangle on the Earth's surface. The three corners of the triangle are the North Pole (point A), the sub-vehicle point (SVP) (point B), and the location of a receiver on the Earth's surface (point C). The SVP is the Earth surface location of the point

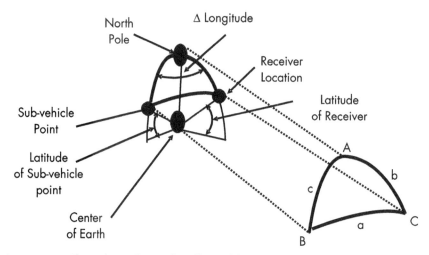

Figure 2.7 This spherical triangle is formed between the North Pole, the SVP, and the receiver location.

directly below the satellite. Angle A is the difference in longitude between the SVP point and the receiver. Angle B is the angular difference between the path from the SVP to the North Pole and the path from the SVP to the receiver. Angle C is the angular difference between the path from the receiver to the North Pole and the path from the receiver to the SVP. Side a is the geocentric angle between the SVP and the receiver. Side b is 90° less the latitude of the receiver. Side c is 90° less the latitude of the SVP.

Figure 2.8 shows a plane triangle in the plane defined by the satellite, the receiver location, and the center of the Earth. The three internal angles are the geocentric angle between the satellite and the receiver location (D), the angle viewed from the satellite between the center of the Earth and the receiver location (E), and the angle viewed from the receiver location between the center of the Earth and the satellite (F). Side d is the direct distance between the satellite and the receiver location. Side e is the radius of the Earth. Side f is the distance from the center of the Earth to the satellite. You may notice that angle D in the plane triangle has the same value as side a in the spherical triangle of Figure 2.7.

You will see these two triangles in later chapters.

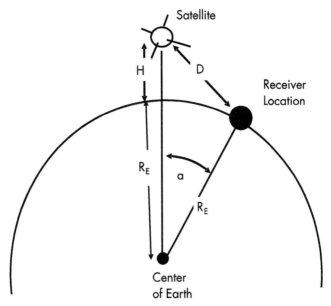

Figure 2.8 The propagation distance between a transmitting satellite and a receiver on the Earth's surface can be calculated from the plane triangle formed by the satellite location, the receiver location, and the center of the Earth.

3

ORBIT MECHANICS

Any satellite orbits around some larger celestial body. In this book, we are concerned with Earth satellites, so each follows an elliptical path in which one of the foci of the ellipse is the center of the Earth as shown in Figure 3.1. To be painfully accurate, the path of the satellite is actually impacted by every other celestial body in the universe; that impact is an inverse function of the range, so the Earth is, by far, the most important influence. We simplify our coverage in this book by only considering the orbiting satellite and the Earth. This is called the two-body problem. Another assumption is that the Earth is a perfect sphere, although you know this is not true. The Earth is proportionally very close to a sphere. The flattening at the poles and the mountain ranges is very small compared to the radius of the Earth.

3.1 ELEMENTS OF AN ELLIPTICAL ORBIT

The location of an Earth satellite relative to the Earth is defined by a series of six numbers that are called the Keplerian ephemeris, because they are credited to the work of Johannes Kepler, a seventeenth-century German mathematician and astronomer. Table 3.1 shows these six numbers and defines them.

a is the semi-major axis shown in Figure 3.1. This is half of the long dimension of the satellite's elliptical orbit. It is also the average distance of the satellite from the center of the Earth. For a circular or-

Table 3.1
Earth Satellite Ephemeris

	Ephemeris Value	Significance
a	Semi-major axis	Size of the orbit
e	Eccentricity	Shape of the orbit
i	Inclination	Tilt of orbit relative to equatorial plane
n	Right ascension of the ascending node	Longitude at which the satellite crosses the equator going north
w	Argument of Perigee	Angle between ascending node and perigee
v	True anomaly	Angle between perigee and the satellite location in the orbit

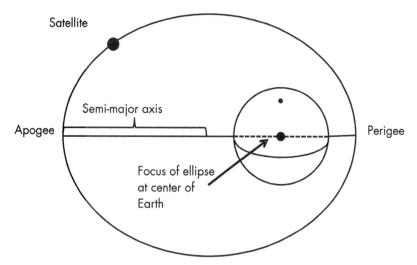

Figure 3.1 An Earth satellite follows an elliptical path with the center of the Earth at one focus of the ellipse.

bit, it is the radius of the orbit. The altitude of a satellite in a circular orbit is constant: the radius of the orbit less the radius of the Earth.

e is the eccentricity of the orbit. This defines the orbit's shape. It is a number between 0 and 1. The distance from the center of the Earth at the satellite's closest approach (the perigee) is $a(1 - e)$ and its minimum altitude is less than this by the radius of the Earth. The maximum distance from the satellite to the center of the Earth (the apogee) is $a(1 + e)$ and its maxim altitude is this distance less the radius of the Earth. For a circular orbit, the eccentricity is zero, and the altitude is constant.

The next four elements of the ephemeris are illustrated in Figure 3.2. All of the angles described are angles as seen from the center of the Earth.

i is the inclination of the orbit relative to the equatorial plane. This determines the maximum latitude covered by the orbit. An equatorial satellite has 0° inclination and a polar orbit has 90° inclination.

n is the right ascension of the ascending node. This is the angle between the longitude of the point at which the satellite crosses the equator going North and the direction of the vernal equinox. The vernal equinox direction is along the line of intersection of the plane of the Earth's orbit around the Sun and the equatorial plane.

w is the argument of perigee. This is the angle between the ascending node and the perigee of the satellite's orbit (in the orbital plane).

v is the true anomaly. This is the angle between the perigee and the satellite location along its orbital path.

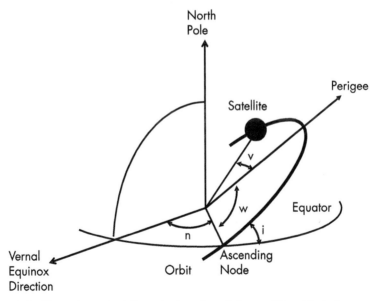

Figure 3.2 The ephemeris defines the location of the satellite with six factors.

3.2 RELATIONSHIP BETWEEN THE SIZE OF AN ORBIT AND ITS PERIOD

Kepler's third law is stated as follows:

$$a^3 = C P^2$$

where a is the semi-major axis of the ellipse of the orbit, C is a constant, and P is the period of the orbit.

Note that the semi-major axis of the ellipse is the radius of a circular orbit. To simplify our lives, we will determine the constant C by considering a circular (i.e., constant altitude) Earth satellite orbit.

Let's look at a satellite that circles the Earth every 1.5 hours and has an altitude of 281.4 km or a radius (from the center of the Earth) of 6,653 km. The constant is:

$$C = a^3/P^2$$

For our 281.4-km-high satellite, C is calculated as:

$$6,653 \text{ km}^3/90 \text{ min}^2 = 36,355,285 \text{ km}^3 \text{ per min}^2$$

This constant value can be used to determine the relationship between the semi-major axis of any Earth satellite and its orbital period. You can subtract the Earth's radius of 6,371 km to find the average altitude of the satellite. For simplicity in this book, we will use circular orbit (i.e., constant altitude) satellites in our examples and problems.

Table 3.2 shows the altitude of a circular Earth satellite versus the period of its orbit for satellites with periods of 1.5 to 9 hours. Figure 3.3 is a graph of the altitude of a circular satellite versus its period. Go straight up from the period (300 minutes, in this case) to the line and then left to the altitude (8,475 km).

Another orbit of particular interest is for the stationary satellite that hovers over a single point on the Earth's surface. Because the Earth rotates 366 times per year (to face the Sun 365 times), the period of the satellite is 23 hours, 56 minutes, and 4.09 seconds (about 1,436 minutes). From Kepler's third law, the radius of this orbit is 42,165.7 km and its altitude is 35,795 km. An additional orbit of interest is that

3.2 RELATIONSHIP BETWEEN THE SIZE OF AN ORBIT AND ITS PERIOD

Table 3.2
Altitude and Semi-Major Axis of Circular Orbits Versus the Satellite Period

p(min)	h(km)	a(km)	p(min)	h(km)	a(km)
90	281	6652	330	9447	15818
105	1001	7372	345	9923	16294
120	1688	8059	360	10392	16763
135	2346	8717	375	10854	17225
150	2980	9351	390	11311	17682
165	3594	9965	405	11761	18132
180	4189	10560	420	12206	18577
195	4768	11139	435	12646	19017
210	5332	11703	450	13081	19452
225	5883	12254	465	13510	19881
240	6422	12793	480	13936	20307
255	6949	13320	495	14357	20728
270	7466	13837	510	14773	21144
285	7974	14345	525	15186	21557
300	8473	14844	540	15595	21966
315	8964	1535	—	—	—

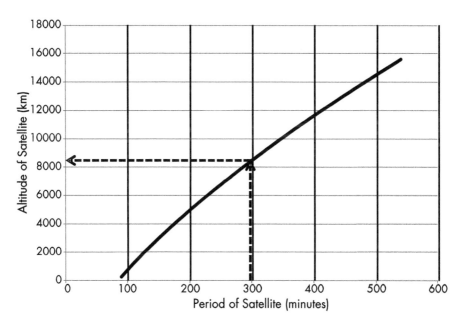

Figure 3.3 The altitude of a circular satellite is a function of its orbital period.

of each of the Global Positioning System (GPS) satellites. They complete two orbits per day. The radii of their 12-hour orbits are 26,612 km (i.e., 20,241-km altitude).

Now we will discuss some general geometrical issues with satellites and their orbits, relative to EW problems. The thrust of this discussion is to show how to determine the range from the satellite to a point on the Earth's surface. This is so we can calculate the link loss, a key element in the effectiveness of jamming or intercept from space.

3.3 SVP

The point on the Earth's surface that is right below the satellite is called the SVP. As shown in Figure 3.4, this point is the intersection of the line from the center of the Earth to the satellite with the surface of the Earth. The latitude of the SVP is the geocentric angle from the equator up to the SVP. The longitude of the SVP is the angle between the longitude line running through the SVP and the longitude line running through Greenwich, England.

Figure 3.4 also shows the definitions of longitude and latitude. Longitude lines are great circle segments; that is, each is on a plane that cuts through the center of the Earth. Latitude lines are not great

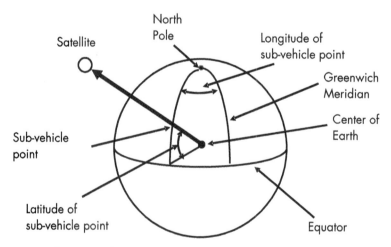

Figure 3.4 The SVP is the intersection of a line from the center of the Earth to the satellite with the Earth's surface.

circles, but the latitude is defined as the angle North or South of the equator to the latitude line.

3.4 AN EARTH SURFACE LOCATION

To simplify the discussion, we will assume that there is a transmitter on the satellite and a receiver on the Earth's surface. Figure 3.5 shows the SVP and the receiver location on the Earth. Each location is defined by its latitude and longitude.

As shown in the figure, there is a spherical triangle with its corners at the North Pole, the SVP, and the receiver location. Using the notation that we discussed in Chapter 2 for spherical triangles, angle A is at the North Pole, angle B is at the SVP, and angle C is at the receiver location. Side c (opposite angle C) is 90° minus the latitude of the SVP. Side b (opposite angle B) is 90° minus the latitude of the receiver location. Side a (opposite angle A) is the geocentric angle between the SVP and the receiver location.

Remember from Chapter 2 that the sides of the spherical triangle are great circle segments and the angles are the intersection angles of the planes containing the sides. The length of a side of a spherical triangle is the angle as seen from the center of the unit sphere (in this case, the center of the Earth).

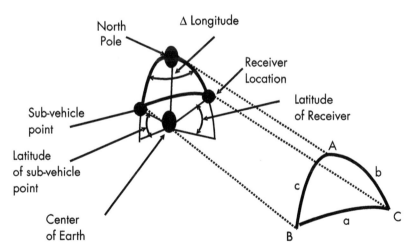

Figure 3.5 A spherical triangle is formed between the North Pole, the SVP, and the receiver location.

We know sides b and c and angle A; we want to determine side a. Thus, we will use the law of cosines for sides, but with our knowns and unknowns placed appropriately in the equation:

$$\cos a = \cos b \cos c + \sin b \sin c \cos A$$

where a, b, c, and A are as defined above.

Now, let's plug in some locations:

Let the SVP be at longitude 200°/latitude 45° and let the receiver be at longitude 230°/latitude 20°. The spherical cosines-for-sides equation is now:

$$\cos a = (\cos c)(\cos b) + (\sin c)(\sin b)(\cos A)$$
$$\cos a = \cos(90-45)\cos(90-20) + \sin(90-45)\sin(90-20)\cos(230-200)$$
$$= \cos(45)\cos(70) + \sin(45)\sin(70)\cos(30)$$
$$= .707 \times .342 + .707 \times .940 \times .866 = .242 + .576 = .818$$

Side a then equals the arc cos of 0.818 = 35.1°.

3.5 THE PROPAGATION DISTANCE

The propagation distance between the transmitter (in the satellite) and the receiver can now be determined from the plane triangle shown in Figure 3.6. Let the satellite be at 1,000-km altitude; the radius of the Earth is 6,371 km. Then the distance between the satellite and the center of the Earth is 7,371 km and side a (from our spherical triangle) is 35.1°.

Figure 3.6 has a plane triangle formed by the satellite, the receiver location, and the center of the Earth. Angle D is side a from Figure 3.5 is 35.1°, side e is 7,371 km, and side f is 6,371 km.

Using the law of cosines for sides formula for plane triangles:

$$d^2 = e^2 + f^2 - 2ef \cos D$$
$$= 7371^2 + 6371^2 - 2(7371)(6371)(.818)$$
$$= 54,331,641 + 40,589,641 - 76,827,609 = 18,093,673$$

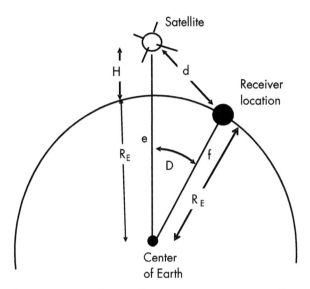

Figure 3.6 The propagation distance between a transmitting satellite and a receiver on the Earth's surface can be calculated from the plane triangle formed by the satellite location, the receiver location, and the center of the Earth.

d (the propagation distance) is the square root of 18,093,673 = 4,254 km.

3.6 EARTH TRACES

In this section, we will discuss some general geometrical issues with Earth satellite orbits. During this discussion, we will be using terms that were defined in Chapter 2. Hopefully, you will forgive me for rounding the numbers during this discussion because we are EW folks rather than orbiteers who must get the numbers exactly right to many decimal places.

3.6.1 Satellite Earth Trace

The Earth trace is the locus of latitude and longitude of the SVP as the satellite moves through its orbit. For low Earth orbits, this determines the moment-to-moment area of the Earth that is seen by the satellite. It also allows us to calculate the look angle and range to the satellite from a specified point on (or above) the Earth at any specified time.

Figure 3.7 is a polar view of the Earth trace of a low-Earth satellite as seen from above the North Pole. Note that the satellite crosses the equator going North at the ascending node and reaches a maximum latitude equal to the inclination of the orbit. This view only traces the North half of the orbit. The other half would be seen in a view from above the South Pole. From the six elements of the ephemeris (defined at the beginning of this chapter), you can calculate the exact location of the satellite at any time. We will be going through that process as we work EW problems in later chapters.

Figure 3.7 shows the Earth traces of two orbits of the satellite. The first orbit Earth trace is shown as a solid line and the next orbit is shown as a dashed line. The orbit is actually affected by mountains and other Earth surface anomalies, but we will assume a perfect spherical Earth in this discussion, so the orbit is assumed to follow a constant path relative t o the center of the Earth as the Earth turns inside of the orbit.

It is easy to become confused by the 3-D geometry when dealing with orbits. There is a simple way to remember which way the Earth spins. The Sun rises in the East once per day. Therefore, the second

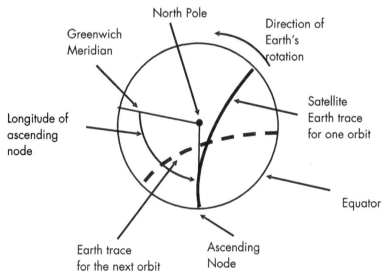

Figure 3.7 The Earth trace of a satellite is the path of the SVP over the Earth's surface. This is a polar view.

orbit will be West of the first orbit by the ratio of the orbital period divided by one sidereal day multiplied by 360°. A sidereal day is 23 hours, 56 minutes and 4.09 seconds (i.e., 1,436 minutes or 86,164 seconds). For example, the Earth trace of a satellite with a 90-minute orbital period will move West by 22.56 longitude degrees for each subsequent orbit.

$$(90/1{,}436) \times 360° = 22.56°$$

Figure 3.8 shows an equatorial view of the Earth trace. In both of Figures 3.7 and 3.8, you can see that the satellite is limited to coverage between the equator and a maximum latitude. The maximum latitude is equal to the inclination of the satellite's orbit relative to the equator.

The Earth area over which a satellite can send or receive signals to and from Earth-based stations during each orbit depends on the altitude of the satellite and the beamwidth and orientation of antennas on the satellite.

3.6.2 Polar Orbit Earth Traces

If a satellite is placed in a polar orbit, its orbit has 90° inclination and will therefore eventually provide complete coverage of the surface of the Earth.

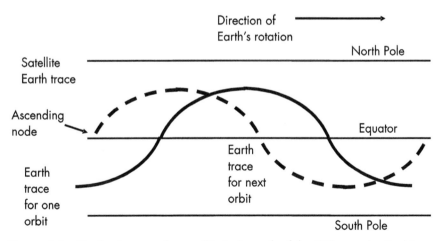

Figure 3.8 The Earth trace of a satellite is the path of the SVP over the Earth's surface. This is an equatorial view.

3.6.3 Synchronous Satellite Earth Traces

A synchronous satellite has an SVP that stays in one location on the Earth's surface. This requires that its orbital period be one sidereal day (i.e., 1,436 minutes). Another requirement for a fixed SVP is that the orbit has 0° inclination. Thus, the SVP must be on the equator. If the synchronous satellite orbit has an inclination, the Earth trace will be a figure-eight as shown in Figure 3.9. This means that an Earth station will see a similar figure-eight variation in the location of the satellite and must therefore track the satellite if a narrow-beam ground antenna is used for transmission or reception.

By Kepler's third law, the altitude of a synchronous satellite must have a semi-major axis of 42,166 km. In a circular orbit, the height of the satellite will be 35,795 km. The maximum range from an Earth surface station to a synchronous satellite with a circular orbit can be determined as shown in Figure 3.10. This diagram is a plane triangle in the plane containing the satellite, the center of the Earth, and an Earth surface station. The Earth station in Figure 3.10 sees the satellite at its local horizon (that is, 0° elevation). The minimum and maximum range values for this satellite to the ground link are 35,795 km and 41,682 km, respectively. The shorter range would apply if the satellite was directly overhead and the maximum range is for the

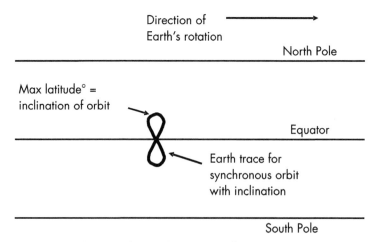

Figure 3.9 The Earth trace of a synchronous satellite is a single spot on the equator unless the orbit has some inclination. Then it is a figure-eight with maximum latitude equal to the orbital inclination.

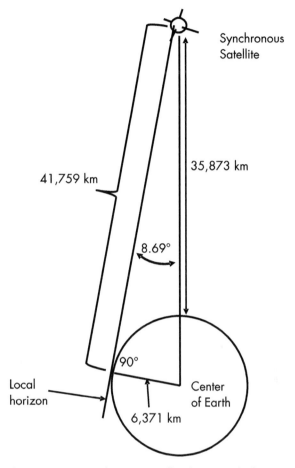

Figure 3.10 The range to a synchronous satellite that is on the horizon is 41,759 km.

satellite on the horizon as shown. This means that the link loss for a 2-GHz signal would be 189.5 dB to 190.9 dB.

Later, we will deal with ranges and look angles for ground and airborne Earth stations for practical EW problems.

You will also note that the angle from the SVP to the signal path tangent to the Earth's surface is 8.69°. If the satellite has an antenna with beamwidth twice this angle (i.e., 17.38°), its beam would cover all of the Earth visible to the satellite. This makes it an Earth-coverage antenna.

3.7 LOCATION OF AN EW THREAT

Since we are starting to talk about space-based EW systems, we need to consider the geometry between the satellite and hostile threat locations. Now we will consider threats on the surface of the Earth. We will just call a hostile transmitter or receiver a threat. An EW system on the satellite will either intercept signals from a threat transmitter or transmit jamming signals to a threat receiver at the considered location.

As shown in Figure 3.11, the location of the threat from the satellite will be defined in terms of the azimuth and elevation of a vector from the satellite that would point at the threat location and the range between the satellite and the threat. You could think of the vector as the pointing information for a satellite antenna aimed at the threat.

3.7.1 Calculating the Look Angles

The azimuth is the angle between true North and the threat location, in a plane at the satellite perpendicular to the vector from the SVP. The elevation is the angle between the SVP and the threat. For the azimuth calculation, we need to consider the spherical triangle in Figure 3.12 defined by the North Pole, the SVP, and the threat location.

For convenience, we define the parts of the triangle in the way that we have in earlier discussion: The uppercase letters are for angles

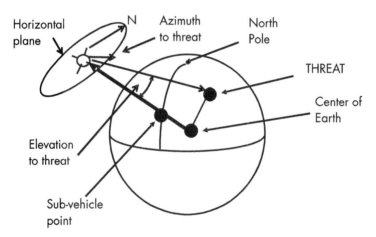

Figure 3.11 The azimuth and elevation angle from the nadir define the direction to a threat from the satellite.

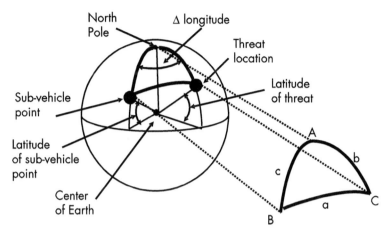

Figure 3.12 A spherical triangle is formed between the North Pole, the SVP, and the threat location.

(i.e., the intersection of two great circle planes through the center of the Earth). The lowercase letters are for sides (i.e., the geocentric angle between points located on one of the great circle planes that form the spherical triangle). Remember that, in a spherical triangle, both the sides and the angles are angles. In this case, A is at the North Pole, B is at the SVP, and C is at the threat location. Side a (opposite angle A) is the path along the Earth's surface from the SVP to the threat. Side b is 90° less the latitude of the threat location. The latitude is the geocentric angle from the equator to some point on the Earth's surface. Side c is 90° less the latitude of the SVP.

3.7.2 Calculating the Azimuth to the Threat

The azimuth to the threat is angle B. (This is what we want to calculate.) Since we start the problem by entering the locations of the SVP and the threat, we know sides b and c and angle A. Angle A is the longitude difference between the SVP and the threat (Δ longitude). Before diving into these trigonometric equations, you may want to review Chapter 2, which defines all of the plane and spherical trigonometric formulas that we will be using over the rest of the book.

First, we calculate side a from the (spherical) law of cosines for sides:

$$\cos a = (\cos b)(\cos c) + (\sin b)(\sin c)(\cos A)$$

Now that we know side a, side b, and angle A, we can find angle B from the (spherical) law of sines:

$$\text{sine of angle } B = (\sin b)(\sin A) / \sin a$$

For example, let's choose a low Earth orbit with a period of 3 hours. From Table 3.2, this satellite will have a semi-major axis of 10,560 km. Let's specify that it has a circular orbit. Then it will be at a constant altitude of 4,189 km.

Our SVP is at 30° North latitude and 100° East longitude. The target that we are considering is on the surface of the Earth at 45° North latitude and 120° East longitude.

The dimensions of the spherical triangle parts in the formula are:

$$\text{Angle } A = 20°$$
$$\text{Side } b = 90° - 45° = 45°$$
$$\text{Side } c = 90° - 30° = 60°$$

Plugging these values into the spherical triangle for sides:

$$\cos a = \cos(45°)\cos(60°) + \sin(45°)\sin(60°)\cos(20°)$$
$$= (0.707 \times 0.5) + (0.707 \times 0.866 \times 0.940) = 0.930$$

So

$$\text{side } a = \arccos(0.930) = 21.57°$$

Now, from the spherical law of sines:

$$\sin(\text{angle } B) = \sin(45°)\sin(20°) / \sin(21.57°)$$
$$= (0.707 \times 0.342) / 0.368 = 0.657$$

So

$$\text{angle } B = \arcsin(0.657) = 41.08°$$

This is the azimuth angle to the target.

3.7.3 Calculating Range and Elevation to the Threat Location

Now consider Figure 3.13. This is a plane triangle in the plane that includes the satellite, the threat, and the center of the Earth. To be consistent with our spherical triangles, we will use uppercase letters for the angles and lowercase letters for the sides. Since this is a plane triangle, the sides are physical lengths, rather than geocentric angles.

E is at the satellite, F is at the threat, and G is at the center of the Earth.

Side e is the radius of the Earth (6,371 km). Side f is the semi-major axis (the radius of the Earth plus the satellite altitude = 10,560 km), angle G is side a from Figure 3.12 (21.57°), and side g is the propagation distance between the satellite and the threat.

The law of cosines for plane triangles is:

$$g^2 = e^2 + f^2 - 2ef \cos(G)$$

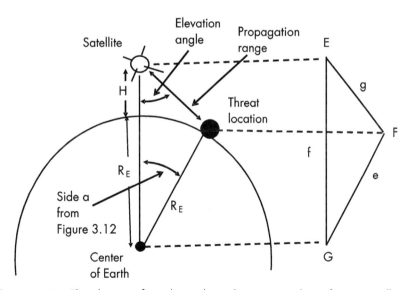

Figure 3.13 The elevation from the nadir and range to a threat from a satellite can be determined from the plane triangle defined by the satellite, the threat, and the center of the Earth.

Plugging in values and solving for g:

$$g = sqrt[6371^2 + 10560^2 - 2\ (6371)(10560)(0.930)] = 5193\ km$$

This is the propagation range from the satellite to the threat.

To find the elevation angle to the threat (angle E) from the plane triangle law of sines

$$\sin E / e = \sin G / g$$

so

$$\sin E = e \sin G / g$$

Plugging in the values: E = arcsin[(6,371)(0.368)/5,193] = arcsin(0.451) = 26.8°

Continuing our discussion of geometrical considerations for space-borne EW systems, we will now calculate the distance over which a satellite can see as a function of its orbital parameters. In Figure 3.14, we did not attempt to draw the satellite altitude to scale; we will handle that in the math. The higher the satellite, the more of the Earth's surface is available for the intercept or jamming of targets.

We have calculated the distance to a target on the ground at a specified latitude and longitude from a satellite that is a specified distance above a specified point on the Earth. We have also calculated the look angles from the satellite to that target. Now we ask the question: Could the satellite actually see that target (i.e., is the target location within the part of the Earth's surface that the satellite can see)? The answer, like many operational answers, is: that depends. In this case, it depends on the specifics of the satellite orbit.

3.8 CALCULATING THE DISTANCE TO THE HORIZON

Consider Figure 3.15. It shows a plane triangle in the plane defined by the satellite, the center of the Earth, and the most distant point on the Earth's surface that the satellite can see. The local horizon plane at the chosen horizon point is tangent to the sphere of the Earth. The vector to this point from the center of the Earth intersects this plane at 90°.

3.8 CALCULATING THE DISTANCE TO THE HORIZON

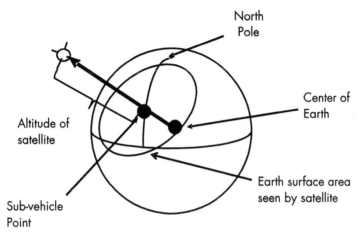

Figure 3.14 The area of the Earth's surface seen by a satellite is a function of the SVP and the altitude of the satellite.

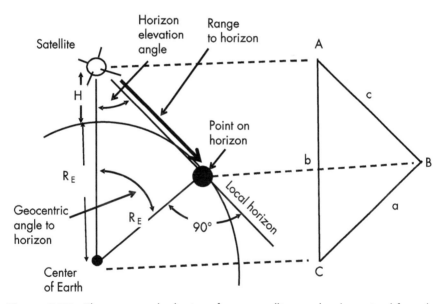

Figure 3.15 The range to the horizon from a satellite can be determined from the plane triangle defined by the satellite, a point on the horizon, and the center of the Earth.

As before, we will pull this triangle out of the diagram and label it in our normal way: that is, uppercase letters for the angles and corresponding lowercase letters for the sides opposite those angles.

Angle A is the elevation angle from the SVP to the horizon. Angle B is 90°. Angle C is the geocentric angle from the satellite to the point on the horizon. Side a is the radius of the Earth. Side b is the distance from the center of the Earth to the satellite (the height of the satellite + the radius of the Earth). Side c is the distance from the satellite to the point on the horizon.

Since this is a right plane triangle, the square of side b is the sum of the squares of sides a and c. So

$$c = sqrt[b^2 - a^2]$$

This is the propagation distance between the satellite and a transmitter or receiver on the horizon.

Now we can use the law of sines for plane triangles to find angle C.

$$\sin C / c = \sin(B)/b \text{ so, } C = \arcsin\left[(c\sin B)/b\right]$$

From angle C, we can determine the Earth surface distance from the SVP to the horizon. The great circle circumference of the Earth is 40,030 km. Thus, we can determine the Earth surface range along a great circle path from the formula:

$$\text{Distance} = 40{,}030 \text{ km (geocentric angle}/360°)$$

In case you want to know it, the elevation angle (from the nadir) to the horizon is 90° less the geocentric angle C we just calculated.

Now let's determine the distance to the horizon for the orbit we are considering. The satellite is in a circular orbit with a 3-hour period, so its altitude is a constant 4,189 km and the radius of its orbit is 10,560 km. The radius of the Earth is 6,371 km.

Plugging these numbers into the above equations:
The link distance to the horizon is:

3.8 CALCULATING THE DISTANCE TO THE HORIZON

$$\text{sqrt}\left[(10560)^2 - (6371)^2\right] = 8422 \text{ km}$$

and the geocentric angle from the satellite to the horizon is:

$$\arcsin\left[((8422)\sin 90°)/10560\right] = \arcsin(0.798) = 52.89°$$

Now we can calculate the Earth surface distance to the horizon (from the law of sines for sides) from:

$$40{,}030 \text{km} (\text{geocentric angle}/360°) = 40{,}030(52.89°/360°) = 5881 \text{ km}$$

In Section 3.7.3, we calculated the straight-line distance to the threat to be 5,193 km. So the link distance is less than the straight-line distance to the horizon (5,193 km versus 8,422 km). Therefore, the satellite can see the target.

The geocentric angle from the SVP to the threat in this problem is 21.57°, so the Earth surface distance is 2,398 km. Thus, the Earth surface distance is less than that to the horizon (2,398 km versus 5,881 km).

The answer, calculated either way, is that the satellite can see the threat location.

3.8.1 Horizon Distances for Circular Orbits

Table 3.3 shows the distance to the horizon for circular satellites with various values of orbit period (in minutes). The first column is the orbital period in minutes, the second column is the altitude of the satellite (h) if it has a circular orbit, the third column shows the semi-major axis (a) (for any orbit shape), the fourth column shows the direct line range (rng) to the horizon in kilometers, and the fifth column shows the Earth surface distance (dist) from the SVP to the horizon in kilometers.

Table 3.3
Height, Semi-Major Axis, and Range to Horizon and Earth Surface Distance to Horizon for Circular Satellites with the Orbital Period Specified

p(min)	h(km)	a(km)	rng(km)	dist(km)	p(min)	h(km)	a(km)	rng(km)	dist(km)
90	281	6652	1914	1859	330	9447	15818	14478	7365
105	1001	7372	3710	3359	345	9923	16294	14997	7447
120	1688	8059	4935	4198	360	10392	16763	15505	7523
135	2346	8717	5950	4785	375	10854	17225	16004	7593
150	2980	9351	6845	5232	390	11311	17682	16494	7658
165	3594	9965	7662	5587	405	11761	18132	16976	7719
180	4189	10560	8422	5880	420	12206	18577	17451	7776
195	4768	11139	9137	6127	435	12646	19017	17918	7830
210	5332	11703	9817	6339	450	13081	19452	18379	7880
225	5883	12254	10467	6523	465	13510	19881	18833	7928
240	6422	12793	11093	6685	480	13936	20307	19281	7973
255	6949	13320	11698	6829	495	14357	20728	19724	8016
270	7466	13837	12284	6958	510	14773	21144	20162	8056
285	7974	14345	12853	7075	525	15186	21557	20594	8095
300	8473	14844	13408	7180	540	15595	21966	21021	8131
315	8964	15335	13949	7277	—	—	—	—	—

4

RADIO PROPAGATION

4.1 INTRODUCTION

The main focus of this chapter is on the basics of radio propagation and how they apply to communications EW. This material is referenced at many other places in the book.

Other material in this chapter relates to the intercept and jamming of normal communication signals. This chapter focuses on radio propagation among target transmitters and receivers within the atmosphere, not the propagation to and from satellites. That comes in Chapter 5. The same EW functions against more complex signals, mainly low probability of intercept signals, will be covered in Chapter 7.

4.2 ONE-WAY LINK

The most dramatic difference between EW against radars and EW against communications is that radars typically use two-way links; that is, the transmitter and receiver are generally (not always) in the same location with transmitted signals repropagating from targets. In communication, the transmitter and the receiver are in different locations. The purpose of communication systems of all types is to take information from one location to another. Thus, communication uses the one-way communication link as shown in Figure 4.1.

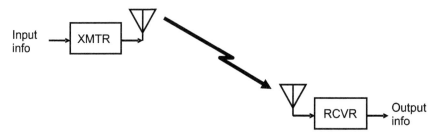

Figure 4.1 A one-way communication link includes a transmitter, a receiver, two antennas, and everything that happens between those antennas.

The one-way link includes a transmitter, a receiver, transmit and receive antennas, and everything that happens to the signal between those two antennas. Figure 4.2 is a diagram that represents the one-way link equation. This diagram is not to scale; it merely shows what happens to the level of a signal as it passes through the link. The ordinate is the signal strength (in dBm) at each point in the link. The transmitted power is the input to the transmit antenna. The antenna gain is shown as positive, although, in practice, any antenna can have positive or negative gain (in decibels). It is important to add that the gain shown here is the antenna gain in the direction of the receiving

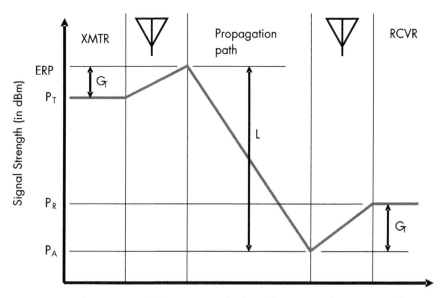

Figure 4.2 The one-way link equation calculates the received power as a function of all other link elements.

4.2 ONE-WAY LINK

antenna. The output of the transmit antenna is called the effective radiated power (ERP) in dBm. Note that the use of dBm units is not technically correct; in fact, the signal at this point is a power density, properly stated in microvolts per meter. However, if we were to place a theoretical ideal isotropic antenna next to the transmit antenna (ignoring the near-field issue), the output of that antenna would be the signal strength in dBm. Using the artifice of this assumed ideal antenna allows us to talk about signal strength through the whole link in dBm without converting units and is thus commonly accepted practice. The formulas to convert back and forth between signal strength in dBm and field density in $\mu v/m$ are:

$$P = -77 + 20\log(E) - 20\log(F)$$

where P is the signal strength arriving at the antenna in dBm, E is the arriving field density in microvolts per meter, and F is the frequency in megahertz.

Conversely, the arriving signal strength can be converted to field density by the formula:

$$E = 10^{[P+77+20\ \log(F)]/20}$$

where E is the field density in microvolts per meter, P is the signal strength in dBm, and F is the frequency in megahertz.

Between the transmit and receive antennas, the signal is attenuated by the propagation loss. We will talk about the various types of propagation loss in detail. The signal arriving at the receiving antenna does not have a commonly used symbol, but we will call it P_A for convenience in some of our later discussions. Because P_A is outside the antenna, it should really be in microvolts per meter, but using the same ideal antenna artifice, we use the units dBm. The receiving antenna gain is shown as positive, although it can be either positive or negative (in decibels) in real-world systems. The gain of the receiving antenna shown here is the gain in the direction of the transmitter.

The output of the receiving antenna is the input to the receiver system in dBm. We call it the received power (P_R). The one-way link equation gives P_R in terms of the other link components. In decibel units, it is:

$$P_R = P_T + G_T - L + G_R$$

where P_R is the received signal power in dBm, P_T is the transmitter output power in dBm, G_T is the transmit antenna gain in decibels, L is the link loss from all causes in decibels, and P_R is the power into the receiver in dBm.

In some literature, the link loss is dealt with as a gain, which is negative (in decibels). When this notation is used, the propagation gain is added in the formula rather than subtracted. In this book, we will consistently refer to loss as a positive number in decibels and therefore subtract loss in link equations.

In linear (i.e., nondecibel) units, this formula is:

$$P_R = (P_T G_T G_R) / L$$

The power terms are in watts, kilowatts, and so forth and must be in the same units. The gains and losses are pure (unitless) ratios. Since the link loss is in the denominator, it is a ratio greater than 1. In subsequent discussions, the loss formulas both in decibels and in linear form will consider loss to be a positive number.

Figures 4.3 and 4.4 show important cases of the use of one-way links in EW. Figure 4.3 shows a communication link and a second link from the transmitter to an intercept receiver. Note that the transmit antenna gain to the desired receiver and to the intercept receiver may be different. Figure 4.4 shows a communication link and a second link from a jammer to the receiver. In this case, the receiving antenna may have different gain toward the desired transmitter and the jammer. Each of the links (in both figures) has the elements shown in the diagram of Figure 4.2.

4.3 PROPAGATION LOSS MODELS

In the description of the link, we clearly separated the transmitting and receiving antenna gains from the link losses. This implies that the link loss is between two unity gain antennas. By definition, an isotropic antenna has unity gain, or 0-dB gain. All of the discussion of link losses in this section will be for propagation losses between isotropic antennas.

4.3 PROPAGATION LOSS MODELS

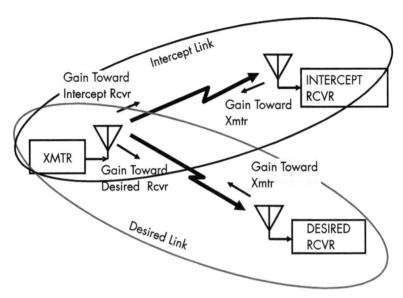

Figure 4.3 When a communication signal is intercepted, there are two links to consider: the transmitter to intercept receiver link and the transmitter to desired receiver link.

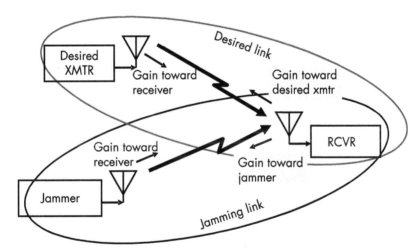

Figure 4.4 When a communication signal is jammed, there is a link from the desired transmitter to the receiver and a link from the jammer to the receiver.

There are a number of widely used propagation models, including the Okumura and Hata models for outdoor propagation and the Saleh and CIRCIM (simulation of indoor radio channel impulse-response models) models for indoor propagation. There is also small-scale fading, which is short-term fluctuation caused by multipath. These models were discussed in an excellent chapter (Chapter 84) in *The Communications Handbook*, published cooperatively by CRC Press and IEEE Press. These detailed models all require computer models of the environment to support analysis of each reflection path in the propagation environment.

Because EW is dynamic by nature, it is common practice not to use these detailed computer analyses, but rather to use three important approximations to determine the appropriate propagation loss models in practical applications. These three models are line of sight (LOS), two-ray, and knife-edge diffraction (KED).

The Communications Handbook also discussed these three propagation models to some extent. Table 4.1 summarizes the conditions under which these three modes are used, and they are described in detail below.

4.3.1 LOS Propagation

LOS propagation loss is also called free space loss or spreading loss. It applies in space and between transmitters and receivers in any other environment in which there are no significant reflectors and the ground is far away in comparison with the signal wavelength (see Figure 4.5).

The formula for LOS loss comes from optics, in which propagation loss is calculated by projecting the transmitting and receiving apertures on a unit sphere with its origin at the transmitter. This is converted to radio frequency propagation by considering the geometry of two isotropic antennas. As shown in Figure 4.6, the isotropic

Table 4.1
Selection of Appropriate Propagation Loss

Clear propagation path	Low frequency, wide beams, near ground	Link longer than Fresnel zone distance	Use two-ray model
		Link shorter than Fresnel zone distance	Use LOS model
	High frequency, narrow beams, far from ground		
Propagation path obstructed by terrain	Calculate additional loss from KED		

4.3 PROPAGATION LOSS MODELS

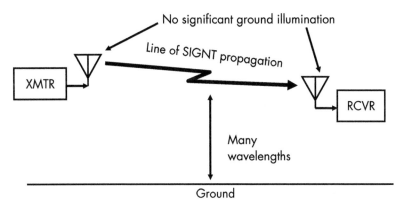

Figure 4.5 If both the transmitter and the receiver are many wavelengths above the ground or if the antenna beams are narrow enough to exclude significant energy to and from the ground, the LOS propagation model is appropriate.

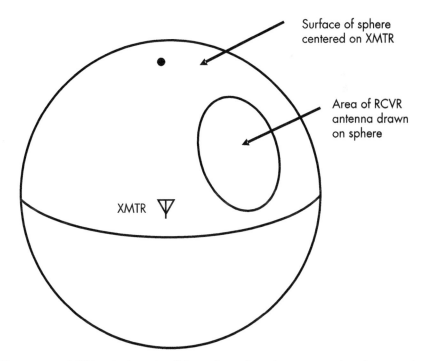

Figure 4.6 LOS loss is the ratio of the surface of a sphere centered on the transmitter with the radius equal to the transmission distance and the effective area of the receiving antenna.

transmitting antenna propagates its signal spherically, with its total energy spread over the surface of the sphere. The sphere expands at the speed of light until its surface touches the receiving antenna. The area of the surface of a sphere is:

$$4\pi R^2$$

where R is in this case the distance from the transmitter to the receiver.

The effective area of the isotropic (i.e., unity gain) receiving antenna is:

$$\lambda^2 / 4\pi$$

where λ is the wavelength of the transmitted signal.

We want the loss to be a number larger than 1, so we can divide the transmitted power by the loss to get the receive power. Thus, we determine the loss ratio by dividing the surface area of the sphere by the area of the receiving antenna:

$$Loss = (4\pi)^2 R^2 / \lambda^2$$

where both the radius and the wavelength are in the same units (typically meters).

Note that some authors treat this as a gain by which the transmitted signal is multiplied. This inverts the right side of the formula.

If we convert from the wavelength to the frequency, the loss formula becomes:

$$Loss = (4\pi)^2 R^2 F^2 / c^2$$

where R is the transmission path distance in meters, F is the transmitted frequency in hertz, and c is the speed of light (3×10^8 m/sec).

Allowing distance to be input in kilometers and frequency in megahertz requires a conversion factor term. Combining terms and converting to decibel form gives the loss in decibels as:

$$L(dB) = 32.44 + 20\log_{10} R + 20\log_{10} F$$

4.3 PROPAGATION LOSS MODELS

where R is the link distance in kilometers and F is the transmit frequency in megahertz. The 32.44 term combines the conversion factors and the c and π terms, converted to decibels. By using this constant, we can input link parameters in the most convenient units.

Alternate forms of this equation change the constant to 36.52 if the distance is in statute miles and to 37.74 if the distance is in nautical miles. The formula is often used in applications to 1-dB accuracy. In this case, the constants are simplified to 32, 37, and 38, respectively.

There is a widely used nomograph that gives the LOS loss in decibels as a function of the distance and the frequency. This is shown in Figure 4.7. To use this nomograph, draw a line between the frequency in megahertz and the link distance in kilometers. Your line crosses the center axis at the LOS loss in decibels. In this figure, the loss at 1 GHz and 10 km is shown as just under 113 dB. Note that the above formula calculates the value at 112.44 dB.

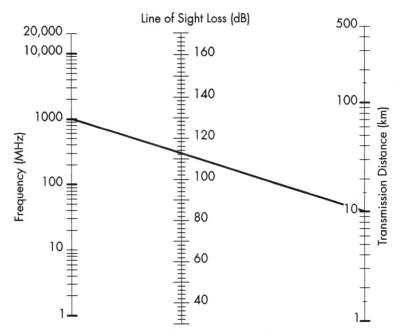

Figure 4.7 A line drawn from the frequency value to the transmission distance value passes through the LOS loss value.

4.3.2 Two-Ray Propagation

When the transmitting and receiving antennas are close to a single dominant reflecting surface (i.e., the ground or water) and the antenna patterns are wide enough to allow significant illumination of that surface, the two-ray propagation model must be considered. As we will see, the transmitted frequency and the actual antenna heights determine whether the two-ray or LOS propagation model applies.

Two-ray propagation is also called 40 log(d) or d^4 attenuation because the loss varies with the fourth power of the link distance. The dominant loss in two-ray propagation is the phase cancellation of the direct wave by the signal reflected from the ground or water as shown in Figure 4.8. The amount of attenuation depends on the link distance and the height of the transmitting and receiving antennas above the ground or water. You will note that (unlike LOS attenuation) there is no frequency term in the two-ray loss expression. In nonlogarithmic form, the two-ray loss is:

$$L = d^4 / \left(h_T^2 \times h_R^2\right)$$

where d is the link distance, h_T is the transmitting antenna height, and h_R is the receiving antenna height. The link distance and antenna heights are all in the same units.

The decibel formula for the two-ray propagation loss is:

$$L = 120 + 40\log(d) - 20\log(h_T) - 20\log(h_R)$$

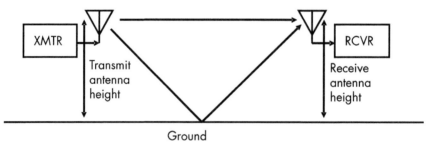

Figure 4.8 In two-ray propagation, the dominant loss effect is the phase cancellation between the direct and reflected signals.

4.3 PROPAGATION LOSS MODELS

where d is the link distance in kilometers, h_T is the transmitting antenna height in meters, and h_R is the receiving antenna height in meters.

Figure 4.9 gives a nomograph for the calculation of two-ray loss. To use this nomograph, first draw a line between the transmitting and receiving antenna heights. Then draw a line from the point at which the first line crosses the center line through the path length to the propagation loss line. In the example, two 10-m-high antennas are 30

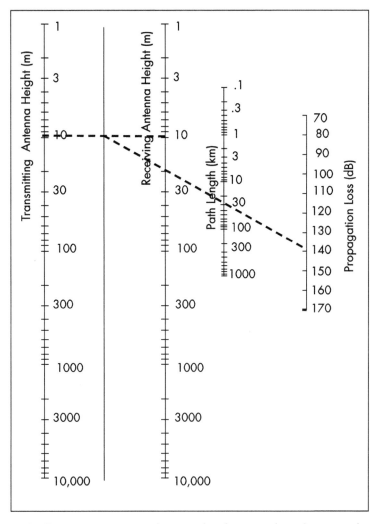

Figure 4.9 Two-ray propagation loss can be determined as shown on this nomograph.

km apart, and the attenuation is a little less than 140 dB. If you calculate the loss from either of the above formulas, you will find that the actual value is 139 dB.

4.3.3 Minimum Antenna Height for Two-Ray Propagation

Figure 4.10 shows the minimum antenna height for two-ray propagation calculations versus transmission frequency. There are five lines on the graph for:

- Transmission over sea water;
- Vertically polarized transmission over good soil;
- Vertically polarized transmission over poor soil;
- Horizontally polarized transmission over poor soil;
- Horizontally polarized transmission over good soil.

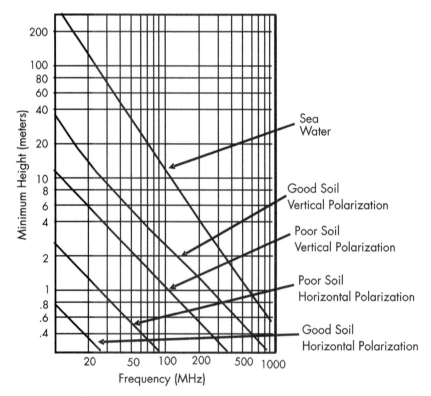

Figure 4.10 If antennas are below the minimum height shown in this graph, use the indicated minimum height in the two-ray propagation loss calculation.

Good soil provides a good ground plane. If either antenna height is less than the minimum shown by the appropriate line in this graph, the minimum antenna height should be substituted for the actual antenna height before completing the two-ray attenuation calculation. Please note that if one antenna is actually at ground level, this chart is highly suspect.

4.3.4 A Note about Very Low Antennas

In the communication theory literature, discussions of very low antennas all seem to be constrained to antenna heights at least a half-wavelength above the ground. A recent, far from complete, test gives some insight into the performance of antennas lower than that. A 400-MHz, vertically polarized, 1-m-high transmitter was moved various distances from a matched receiver while the receiver was lowered from 1m high to the ground. Over level, dry ground, the received power reduced by 24 dB when the receiving antenna was at the ground. With a 1-m-deep ditch across the transmission path (near the receiver), this loss was reduced to 9 dB.

4.3.5 Fresnel Zone

As mentioned above, signals propagated near the ground or water can experience either LOS or two-ray propagation loss, depending on the antenna heights and the transmission frequency. The Fresnel zone distance is the distance from the transmitter at which the phase cancellation becomes dominant over the spreading loss. As shown in Figure 4.11, if the receiver is less than the Fresnel zone distance from the transmitter, LOS propagation takes place. If the receiver is farther than the Fresnel zone distance from the transmitter, two-ray propagation applies. In either case, the applicable propagation applies over the whole link distance.

The Fresnel zone distance is calculated from the following formula:

$$FZ = 4\pi h_T h_R / \lambda$$

where FZ is the Fresnel zone distance in meters, h_T is the transmitting antenna height in meters, h_R is the receiving antenna height in meters, and λ is the transmission wavelength in meters.

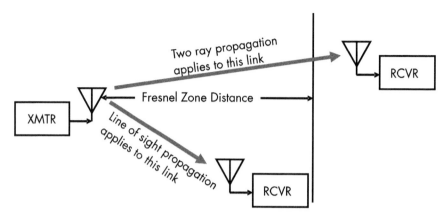

Figure 4.11 If the link is shorter than the Fresnel zone distance, it uses LOS propagation. If it is longer than the Fresnel zone distance, it uses two-ray propagation.

Note that several different formulas for Fresnel zone are found in literature. The formulas are similar, but give different answers and discontinuities. This one is chosen because it yields the distance at which LOS and two-ray attenuations are equal and thus provides a realistic result. A more convenient form of this equation is:

$$FZ = [h_T \times h_R \times F]/24{,}000$$

where FZ is the Fresnel zone distance in kilometers, h_T is the transmitting antenna height in meters, h_R is the receiving antenna height in meters, and F is the transmission frequency in megahertz.

4.3.6 Complex Reflection Environment

In locations with very complex reflections, for example, when transmitting down a valley as shown in Figure 4.12, it has been suggested in the literature that the LOS propagation loss model will give a more accurate answer than the two-ray propagation model. This agrees with the author's field experience.

4.3.7 KED

Non-LOS propagation over a mountain or ridge line is usually estimated as though it were propagation over a knife edge. This is a very common practice and many EW professionals report that the actual

4.3 PROPAGATION LOSS MODELS

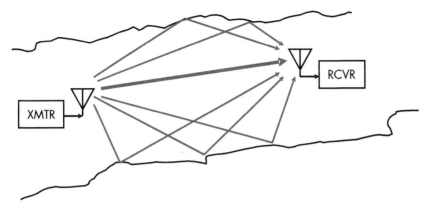

Figure 4.12 In a very complex reflection environment, like transmission down a valley, the actual propagation loss can be expected to be closer to LOS than two-ray propagation.

losses experienced in terrain closely approximate those estimated by equivalent KED estimation.

The KED attenuation is added to the LOS loss as it would be if the knife edge (or ridge line) were not present. Note that the LOS loss rather than the two-ray loss is added to the KED loss even though the link distance is longer than the Fresnel zone (see Figure 4.13).

The geometry of the link over a knife edge is shown in Figure 4.14. H is the distance from the top of the knife edge to the LOS as though the knife edge were not present. The distance from the transmitter to the knife edge is called d_1, and the distance from the knife edge to the receiver is called d_2. For KED to take place, d_2 must be at least equal to d_1. If the receiver is closer to the knife edge than the transmitter, it is in a blind zone in which only tropospheric scattering (with significant losses) provides link connection.

As shown in Figure 4.15, the knife edge causes loss even if the LOS passes above the peak, unless the LOS path passes several wavelengths above. Thus, the height value H can be the distance either above or below the knife edge.

Figure 4.16 is a KED calculation nomograph. The left-hand scale is a distance value d, which is calculated by the following formula:

$$d = \left[sqrt\ 2/(1 + d_1/d_2) \right] d_1$$

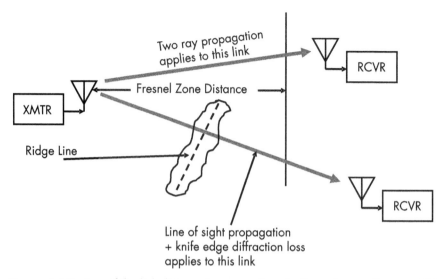

Figure 4.13 Even if the link distance is greater than the Fresnel zone distance, LOS propagation and the KED factor apply if there is an intervening ridge line.

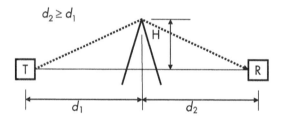

Figure 4.14 The KED geometry is set by the distance to the knife edge, the distance past the knife edge, and the height of the knife edge relative to the LOS path if there were no knife edge.

where d_1 is the horizontal distance from the transmitter to the ridge line and d_2 is the horizontal distance from the ridge line to the receiver.

Table 4.2 shows some calculated values of d.

If you skip this step and just set $d = d_1$, the KED attenuation estimation accuracy will only be reduced by about 1.5 dB.

Returning to Figure 4.16, the line from d (in kilometers) passes through the value of H (in meters). At this point, we do not care whether H is the distance above or below the knife edge. Extend this line to the center index line.

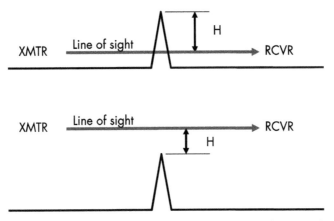

Figure 4.15 The LOS path can pass above or below the knife edge. If it is not too far above, KED loss will still occur.

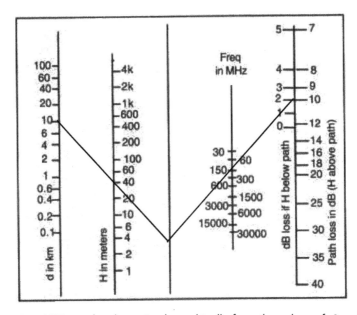

Figure 4.16 KED can be determined graphically from the values of d and H and the frequency.

Another line passes from the intersection of the first line with the center index through the transmission frequency (in megahertz) to the right-hand scale, which gives the KED attenuation. At this point, we specify whether H was above or below the knife

Table 4.2
Values of d

	D
$d_2 = d_1$	$0.707\ d_1$
$d_2 = 2\ d_1$	$0.943\ d_1$
$d_2 = 2.41\ d_1$	d_1
$d_2 = 5\ d_1$	$1.178\ d_1$
$d_2 \gg d_1$	$1.414\ d_1$

edge. If H is the distance above the knife edge, the KED attenuation is read on the left-hand scale. If H is the distance below the knife edge, the KED attenuation is read on the right-hand scale.

Consider an example (which is drawn onto the nomograph): d_1 is 10 km, d_2 is 24.1 km, and the LOS path passes 45m below the knife edge.

d is 10 km and H is 45m. The frequency is 150 MHz. If the LOS path were 45m above the knife edge, the KED attenuation would have been 2 dB. However, since the LOS path is below the knife edge, the KED attenuation is 10 dB.

The total link loss is then the LOS loss without the knife edge and the KED attenuation:

$$\text{LOS loss} = 32.44 + 20\log(d_1 + d_2) + 20\log\ (\text{frequency in MHz})$$
$$= 32.44 + 20\log(34.1) + 20\log(150) = 32.44 + 30.66 + 43.52$$
$$= \text{approximately } 106.6 \text{ dB}$$

So the total link loss is 106.6 + 10 = 116.6 dB.

4.3.8 Atmospheric Loss

Atmospheric loss is caused by the absorption of radio energy (mainly by oxygen and water vapor) as signals pass through the Earth's atmosphere. The combined losses per kilometer are shown as a function of frequency in Figure 4.17. Note that this curve is for attenuation horizontally through the troposphere near the Earth's surface.

To use this chart, enter at the transmission frequency along the bottom of the chart, move straight up to the curve, and read the

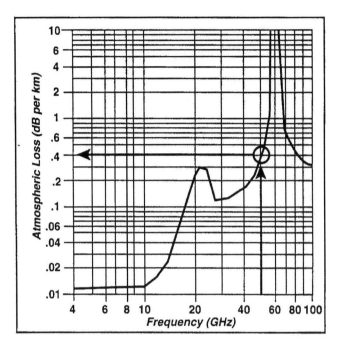

Figure 4.17 Atmospheric attenuation.

loss per unit distance on the left scale. Assuming that the whole transmission path is in the atmosphere, the total atmospheric loss is the loss per unit distance multiplied by the link distance. For example, at 47 GHz, the atmospheric loss is about 0.4 dB/km, so a 10-km link at 47 GHz would have 4 dB of atmospheric loss.

Notice that there is a very strong peak in atmospheric attenuation at about 60 GHz, which would make this a very poor choice for an Earth transmission link, but excellent for transmitting from satellite to satellite if you do not want anyone on Earth to listen in.

Another thing to notice about this chart is that the atmospheric attenuation gets very small at low frequencies. Below microwave frequencies, it is usually acceptable to ignore the atmospheric attenuation as negligible, particularly when making calculations to 1-dB accuracy.

4.3.9 Rain and Fog Attenuation

Radio signals are attenuated as they pass through rain and fog in excess of the normally accepted atmospheric attenuation levels.

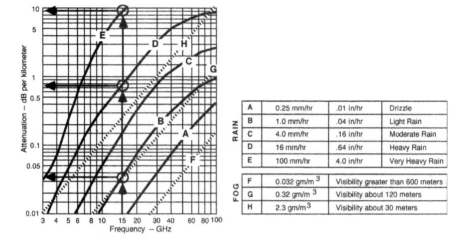

Figure 4.18 Rain loss is a function of frequency and the rate of rainfall or the density of the fog.

Figure 4.18 shows the amount of additional attenuation per unit distance (in kilometers) that is caused by various densities of rainfall and fog. This chart is representative of several found in various references. The table included in the figure defines the rainfall rate and fog visibility distances for various levels of rain and fog.

To use the chart in Figure 4.18, enter it from the bottom at the link operating frequency, go up to the line corresponding to the rainfall or fog density, and then go left to the attenuation per kilometer. The additional attenuation is this number multiplied by the path length (in kilometers) over which the link is subject to that rain or fog density.

5

RADIO PROPAGATION IN SPACE

Radio propagation within the atmosphere was covered in Chapter 4, but when radio transmission is to or from an Earth satellite, there are some special considerations. Some have to do with the nature of space and some are related to the geometry of the link. The received power at the receiver is given by the following formula:

$$P_R = ERP - L + G_R$$

where P_R is the power into the receiver, ERP is the effective radiated power from the transmitting antenna, L is the loss between the transmitting and receiving antennas, and G_R is the gain of the receiving antenna.

This chapter focuses on the losses part of this equation, most specifically on the losses that are peculiar to links that pass through both space and the Earth's atmosphere.

The total path loss to or from a satellite includes the LOS loss, atmospheric loss, antenna misalignment loss, polarization loss, and rain loss.

5.1 LOS LOSS

The primary component of link loss to or from a satellite is associated with the extremely long range. Since the satellite is in space, this loss is the space loss, also called the LOS loss, determined from the formula:

$$L = 32.44 + 20\log(d) + 20\log(F)$$

where L is the LOS loss in decibels, d is the transmitter to receiver distance in kilometers, and F is the transmission frequency in megahertz.

Note that this is the loss between two isotropic (i.e., 0-dB gain) antennas. The gains and losses associated with the transmitting and receiving antennas are considered separately.

On large satellites, the link antennas are typically parabolic dishes with large gain and narrow beams. The gain can be calculated from the operating frequency and diameter by the following formula:

$$G = -42.2 + 20\log(D) + 20\log(F)$$

where G is the antenna gain (in decibels), D is the reflector diameter (in meters), F is the frequency (in megahertz), and the antenna efficiency is 55%.

The antenna beamwidth can be calculated from the formula:

$$BW = \mathrm{Antilog}\left[(86.8 - 20\log(D) - 20\log(F))/20\right]$$

where BW is the 3-dB beamwidth (in degrees), D is the reflector diameter (in meters), and F is frequency (in megahertz).

On smaller satellites, the antennas can be spirals or dipoles.

If the distance from the satellite to the ground station is, for example, 1,000 km and the link is operating at 15 GHz, the space loss is 175.96 dB.

5.1.1 Atmospheric Loss

Figure 5.1 shows the atmospheric loss per kilometer for signals transmitted within the atmosphere. However, when a transmission is to or from a satellite, we use Figure 5.2. This is the loss through the whole atmosphere as a function of frequency and the elevation angle of the path to the satellite relative to the local horizon. For example, for a 3-GHz signal, the atmospheric loss is negligible at 30° elevation, but about 2.2 dB at the horizon as shown in Figure 5.3.

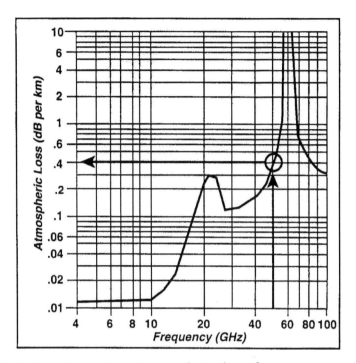

Figure 5.1 Atmospheric attenuation near the Earth's surface.

5.2 ANTENNA MISALIGNMENT

Where narrow-beam antennas are used in satellite links, the alignment of the transmitting and/or receiving antenna will cause a link loss because of a reduction in the effective antenna gain as shown in Figure 5.4. For small antenna misalignments, the gain reduction can be calculated from the formula:

$$\Delta G = 12\, (\theta / \alpha)^2$$

where ΔG is the reduction from boresight gain (in decibels), α is the 3-dB beamwidth (in degrees), and θ is the offset angle (in degrees).

For example, if the transmitting antenna has 5° 3-dB beamwidth and is misaligned by 1°, but the receiving antenna boresight is aimed directly at the transmitting antenna, the received signal strength will be reduced by the transmitting antenna gain reduction factor:

$$\Delta G = 12(1/5)^2 = 0.48 \text{ dB}$$

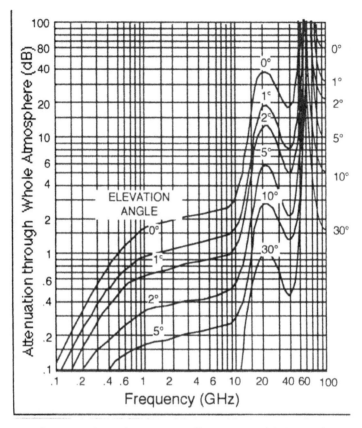

Figure 5.2 The atmospheric loss in a satellite to ground link is a function of frequency and the satellite elevation angle.

Note that the alignment of both the transmitting and receiving antennas must be considered when figuring total link losses.

5.3 POLARIZATION LOSS

Another source of loss is from antenna polarization incompatibility. If the polarization of the signal arriving at the receiving antenna is the same as the polarization of that antenna, there is no polarization loss. If not, there is a polarization loss. A complication for links between satellites and Earth stations is that passage through the atmosphere can rotate the polarization. If linear polarizations are used in link antennas, this rotation can cause several decibels of loss. Therefore,

5.3 POLARIZATION LOSS

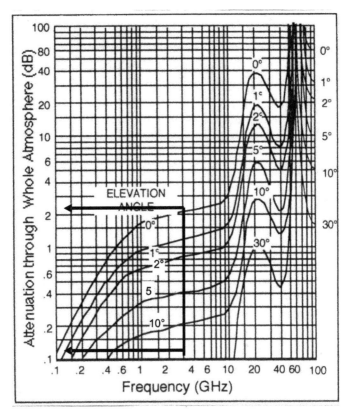

Figure 5.3 The atmospheric loss through the whole atmosphere at 5 GHz with a 0° elevation is a little over 2 dB but is negligible at 30° elevation.

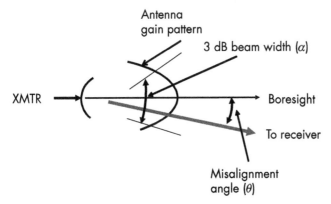

Figure 5.4 For small misalignment errors, the antenna gain is reduced by a function of the angular error and the antenna 3-dB beamwidth.

circular polarization is usually used in space to Earth station links; the rotation of polarization through the atmosphere then causes no link loss. Circular polarization can be right-hand or left-hand. If the transmitting and receiving antennas are not the same, there can be significant mismatch loss. For small EW antennas, this can be on the order of 10 dB, but for typical (larger) spacecraft and ground station antennas, the right/left mismatch error can be more than 30 dB. This allows the link frequency to be used twice; for example, the uplinks and downlinks for a satellite can use the same antenna at the same time if the two links have opposite-sense circular polarization.

Another polarization consideration is that small ground or airborne antennas sometimes have linear polarization, while the spacecraft antennas are circularly polarized. When one end of the link has any linear polarization while the other end of the link has either right-hand or left-hand circular polarization, there is a 3-dB polarization loss. Figure 5.5 summarizes the polarization mismatch issue. As we discuss the links between the satellite and hostile transmitting or receiving antennas, there will be more detailed discussion of polarizations.

5.4 RAIN LOSS

Figure 5.6 shows the link loss per kilometer caused by rain. To use this chart, start with the frequency on the abscissa. Move vertically to the line for the rain rate anticipated and then move left to the ordinate, which gives the rain loss per kilometer. The table included in Figure 5.6 identifies the curve on the loss graph for each of several rainfall rates and fog densities. The loss per kilometer is then multiplied by the distance over which the link passes through the rain.

For space links, there is another complication. Out in space, there is no rain. Thus, the rain loss applies only between the altitude at which the rain starts and the altitude of the Earth station. Figure 5.7 shows the altitude of the 0° isotherm versus latitude and probability. Rain falls from the altitude at which the atmosphere is at 0° centigrade (i.e., the 0° isotherm). Above this altitude, the rain is frozen, and ice causes negligible attenuation. Consider the elevation angle of the satellite from the Earth terminal. We can

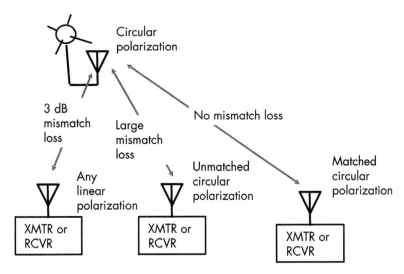

Figure 5.5 The satellite link polarization mismatch loss is zero for matched circular polarized antennas, large for mismatched circular polarized antennas, and 3 dB for circular to linear polarized antennas.

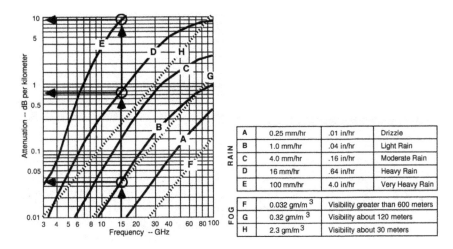

Figure 5.6 Rain loss is a function of frequency and the rate of rainfall.

calculate the length of the path through the rain as the height of the 0° isotherm divided by the sine of this angle as shown in Figure 5.8. Note that if the link is between a satellite and an aircraft, the

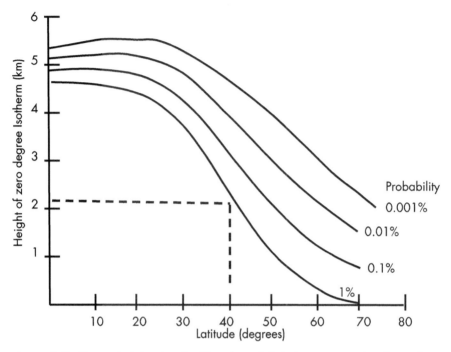

Figure 5.7 The probability that the 0° isotherm falls below a given altitude is a function of the latitude.

Earth terminal is at the altitude of the aircraft. Since rain attenuation is considered probabilistically rather than calculated moment to moment, the chart in Figure 5.7 (the probability of the height of the 0° isotherm versus latitude) is very helpful. If, for example, we are considering the 99% reliability of a link performance level, we would start at the latitude of the Earth station and go up to the 1% line and then left to the ordinate to find the expected height of the 0° isotherm. Let's take an example: We are at 40° latitude. The 1% line is at about 2-km altitude. The satellite is 30° above the local horizon. The sine of 30° is 0.5. The range through the rain is thus 2 km/0.5 = 4 km. If there is heavy rain and the link is at 15 GHz, the rain loss (from Figure 5.6) will be 0.73 dB/km multiplied by 4 km, which is 2.9 dB. If light rain was present, the loss per kilometer would be 0.033 dB/km, so the rain loss would be 0.3 dB over 10 km.

Figure 5.9 shows that the elevation of the satellite above the local horizon can be determined from the plane triangle defined

5.4 RAIN LOSS

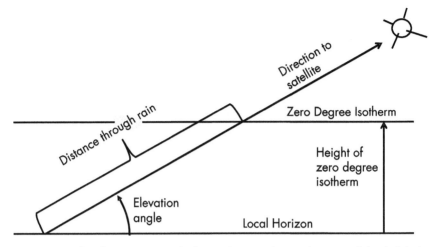

Figure 5.8 The distance over which rain loss applies is the part of the link below the 0° isotherm.

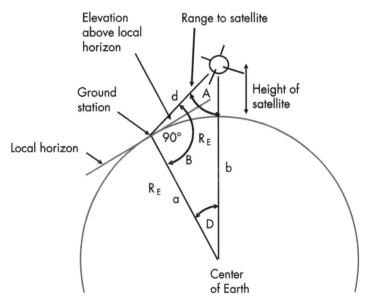

Figure 5.9 The elevation to an Earth satellite can be calculated from a plane triangle formed by the satellite, a ground station, and the center of the Earth.

by the ground station location, the satellite, and the center of the Earth.

As an example, let's use a satellite and ground station location from Chapter 3.

The satellite is at 1,000-km altitude in a circular orbit. This means that side a in Figure 5.9 is 6,371 km and side b is 7,371 km. The SVP for the satellite is 200° East longitude, 45° North latitude. The ground station is at 230° East longitude and 20° North latitude.

From the calculations in Sections 3.4 and 3.5, we know that the geocentric angle between the SVP and the ground station (which is angle D in Figure 5.9) is 35.1°. The link distance from the satellite to the ground station (side d in Figure 5.9) is 4,254 km.

Using the law of sines for plane triangles:

Angle B =
$\arcsin[(b \sin D) d = [(7371 \times .5750)/4254] = \arcsin[0.9963]\ 85,1°$ or 94.9°

Since angle B is greater than 90°, the angle is 94.9°.

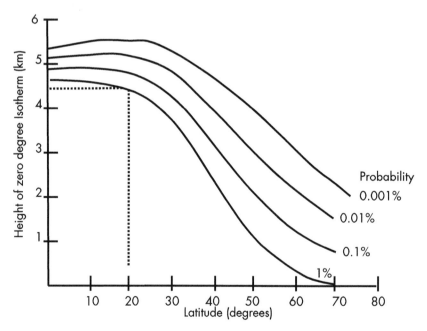

Figure 5.10 The 1% 0° isotherm at 20° latitude is at 4.4 km.

5.4 RAIN LOSS

The horizon line makes a 90° angle with side a in the figure, so the satellite is 4.9° above the horizon from the ground station.

Figure 5.10 shows that the 1% zero degree isotherm altitude is 4.4 km at latitude 20 degrees. So the distance that the link travels through the rain is:

$$4.4 \text{ km} / \sin 4.9° = 4.4 \text{ km} / .089854 = 49 \text{ km}$$

Referring back to Figure 5.6, if the link is at 15 GHz and there is heavy rain, the rain loss would be 0.73 dB/km. Thus, the rain loss is 49 km × 0.073 dB/km = 35.8 dB.

If the link is at a lower frequency (let's choose 5 GHz), the loss per kilometer is only 0.043 dB/km, so the rain loss is only 2.1 dB.

6

SATELLITE LINKS

Satellites are, by their nature, remote from the ground and must be connected by links.

6.1 LINK GEOMETRY

This geometry applies to all satellite links: both uplinks and downlinks. There are differences in uplinks and downlinks that will be dealt with later in this chapter.

The spatial and geometric relationships discussed here will be the basis of later discussions of each type of link, so it is important to be able to picture them. Even after many years, I am required to go through the thought process of picturing standing on the satellite and looking at the ground station and moving a hand through each of the several angles to avoid confusion.

It is important to understand that the figures in this chapter are not drawn with the actual angles involved. They are drawn so that there is room in the figures for the labels.

Figure 6.1 shows the satellite's SVP. As discussed in Chapter 3, this is the intersection of a line between the satellite and the center of the Earth with the surface of the Earth. We are assuming here that the Earth is a perfect sphere. The location of the SVP is defined by its latitude and longitude. To review from Chapter 3, the latitude is the geocentric angle from the Equator to the SVP. The longitude is the geocentric angle between the plane containing the North Pole, the center

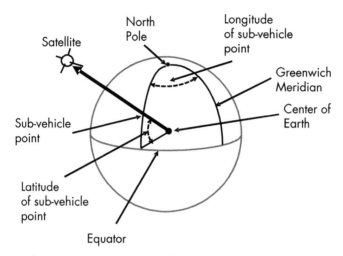

Figure 6.1 The SVP is the intersection of a line from the center of the Earth to the satellite with the Earth's surface.

of the Earth, and the SVP and the plane that includes the North Pole, the center of the Earth, and Greenwich, England. This latter is called the Greenwich meridian. The longitude of anything on the Earth's surface is defined relative to this meridian.

A great circle is in a plane that passes through the center of the Earth. Note that latitude lines do not form great circle paths on the Earth's surface. Picture the planes defined by latitude lines as slices through the Earth parallel to the Equator. Since these slices do not pass through the center of the Earth (except for the Equator itself), they do not contain great circles.

Figure 6.2 shows the location of the ground station. At this point, ground station is taken to mean the Earth surface end of a link. It can be the satellite's ground control station, a hostile receiver site, or a hostile transmitter site. Uplinks go from this ground station to the satellite and downlinks go from the satellite to the ground station. Its latitude is the geocentric angle between the Equator and the ground station location. Its longitude is the geocentric angle between the plane containing the North Pole, the center of the Earth, and the ground station and the Greenwich meridian. Since they are in planes that pass through the center of the Earth, longitude lines are great circles.

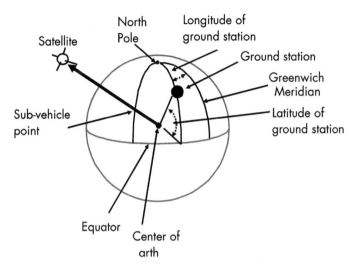

Figure 6.2 The ground station in this diagram can be the satellite's control station or a hostile transmitter or receiver.

Figure 6.3 deals with the relative location of these two points and the satellite. The elevation of the ground station as seen from the satellite is the angle above the center of the Earth to the ground station. The uplink from the ground station to the satellite is also shown in this figure.

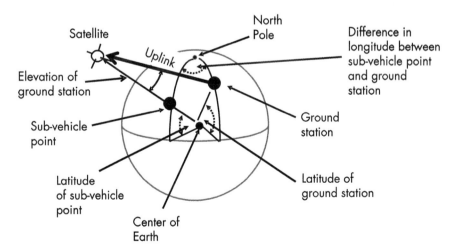

Figure 6.3 The relative location of the SVP and the ground station is a function of the two latitudes and the difference in longitude.

6.1.1 Looking Down

The direction from the satellite to the ground station defines the way that a directional satellite antenna would need to be aimed to direct it at the ground station as shown in Figure 6.4. If we are discussing uplinks, this would be a receiving antenna. If we are discussing downlinks, it would be a transmitting antenna. It would need to be aimed at the proper azimuth and elevation from the satellite's location as shown in the figure. The elevation is the angle between the center of the Earth and the ground station (as seen from the satellite). The azimuth is defined in a plane perpendicular to a line between the satellite and the center of the Earth. The azimuth is the angle in this plane between the North Pole and the ground station.

Now consider the spherical triangle of Figure 6.5.

Angle A is the difference in longitude between the SVP and the ground station.

Side b is 90° minus the latitude of the ground station.

Side c is 90° minus the latitude of the SVP.

With these three values, we can determine the third side, side a, using the spherical law of cosines for sides:

$$\cos a = (\cos b)(\cos c) + (\sin b)(\sin c)(\cos A)$$

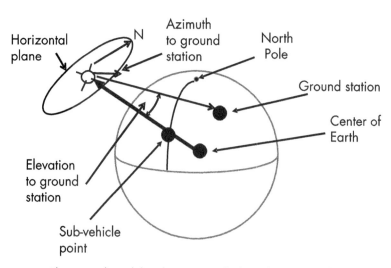

Figure 6.4 The azimuth and the elevation angle from the nadir define the direction to a ground station from the satellite.

6.1 LINK GEOMETRY

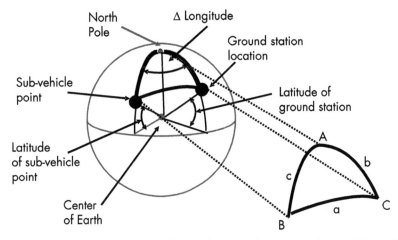

Figure 6.5 A spherical triangle is formed between the North Pole, the SVP, and the ground station.

The azimuth to the ground station is angle B. We can find this by reorganizing the spherical law of cosines for sides:

$$B = \arccos\left[(\cos b - (\cos a)(\cos c))/((\sin a)(\sin c))\right]$$

Figure 6.6 is a plane triangle formed by the satellite, the ground station, and the center of the Earth. To avoid confusion, we have given this plane triangle new dimension names as shown in the figure.

Side e is the radius of the Earth

Side f is the height of the satellite + the radius of the Earth.

Angle G in Figure 6.6 is the same as side a in Figure 6.5. We can use the law of cosines for sides for plane triangles:

$$(\text{Side } g)^2 = f^2 + e^2 - 2fg\cos G$$

to determine the distance from the satellite to the ground station. So:

$$\text{Side } g = \text{SQRT}\left[f^2 + e^2 - 2fe\cos G\right]$$

The elevation angle to the ground station from the satellite is angle E.

From the law of sines for plane triangles:

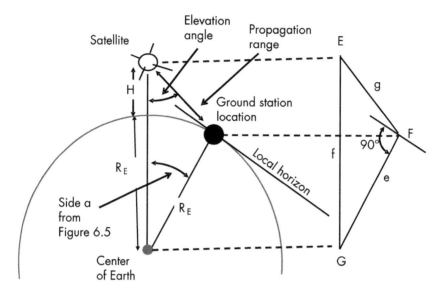

Figure 6.6 The elevation from the nadir and the range to the ground station from a satellite can be determined from the plane triangle defined by the satellite, the ground station, and the center of the Earth.

$$\sin G / g = \sin F / f = \sin E / e$$
$$\text{Angle } E = \arcsin\left[e(\sin G / g)\right]$$

6.1.2 Plug in Some Orbit Numbers

Let's put the SVP at longitude 20° East of the Greenwich meridian and 45° North latitude. The ground station location is at 28° East longitude, 55° North latitude.

The satellite has a 1,000-m-high circular orbit and the radius of the Earth is 6,371 km.

For the spherical triangle in Figure 6.5:

Angle $A = 8°$
Side b is $90° - 55° = 35°$
Side c is $90° - 45° = 45°$

We find side a from the spherical law of cosines for sides:

6.1 LINK GEOMETRY

$$\text{Side } a = \arccos[(\cos 35°)(\cos 45°) + (\sin 35°)(\sin 45°)(\cos 8°)]$$
$$= \arccos[(.819)(.707) + (.574)(.707)(.990)]$$
$$= \arccos[.579 + .402] = \arccos[.981] = 11.2°$$

Angle B in Figure 6.5 is the azimuth from the satellite to the ground station.

$$B = \arccos\left[(\cos b - (\cos a)(\cos c))/((\sin a)(\sin c))\right]$$
$$B = \arccos[((\cos 35°) - (\cos 11.2°)(\cos 45°)/((\sin 11.2°)(\sin 45°))]$$
$$B = \arccos\left[((819) - (.981)(.707))/((.1942)(.707))\right]$$
$$= \arccos\left[(.819 - .6936)/.1373\right] = \arccos[.9133] = 24.0°$$

Now consider the plane triangle in Figure 6.6:
Side f is 6,371 km + 1,000 km = 7,371 km.
Side e is 6,371 km.
Angle G is the same as side a in Figure 6.5 = 11.2°.

$$\text{Side } g = \text{SQRT }[(7,371)^2 + (6,371)^2 - 2(7,371)(6,371)\cos(11.2°)]$$
$$= \text{SQRT}[54,331,641 + 40,589,641 - 46,066,285]$$
$$= \text{SQRT}[48,854,997] = 6990 \text{ km}$$

This is the downlink distance.
Now the elevation angle from the satellite to the ground station is:

$$\text{Angle } E = \arcsin\left[e(\sin G / g)\right]$$
$$\text{Angle } E = \arcsin[6371(\sin(11.2°)/6990]$$
$$= \arcsin[(6371)(.1942)/6990] = \arcsin[.1770]$$

6.1.3 Looking Up

Now let's look the other way at the satellite to ground station link. We want to know the azimuth and elevation of the satellite as seen from the ground station. Look back at the spherical triangle in Figure 6.5. Angle C is the azimuth to the satellite as seen from the ground station. We are using the same orbital values that we used in Section 6.1.1.

Since we know angle A (8°), side a (11.2°), and side c (45°), we can find angle C from the spherical law of sines: sinC/sin c = sinA/sin a, but we know from the physical situation (the SVP is South of the ground station (45° latitude versus 55° latitude) and the longitude difference is small (8°)) that angle C is greater than 90°.

$$C = \arcsin\left[\sin c(\sin A / \sin a)\right]$$
$$= \arcsin\left[\sin(45°)(\sin(8°) / \sin(11.2°)\right]$$
$$= \arcsin\left[.7071(.1392/.1942)\right]$$
$$= \arcsin[.5068] = 149.5°$$

However, the satellite is West of the SVP, so the azimuth is 360° − 149.5° = 210.5°.

The elevation of the satellite from the ground station is the angle above the local horizon. From Figure 6.6, the local horizon is shown to make a 90° angle to side e at the ground station location. This means that the elevation of the satellite from the ground station is angle F minus 90°.

Since this is a plane triangle, the sum of the three angles is 180°, and we know that angle E is 10.2° and angle G is 11.2°. Thus, angle F is:

$$180° - 10.2° - 11.2° = 158.6°$$

The elevation above the local horizon is 158.6° − 90° = 68.6°.

6.2 UPLINKS

Uplinks have transmitters on or near the Earth's surface and receivers in the satellite. In this section, we will discuss uplinks to the satellite from several different types of emitters:

6.2 UPLINKS

- The ground station controlling the satellite or its payload;
- A hostile emitter that the satellite payload is intercepting;
- A ground-based jammer that is jamming the satellite uplink.

All of these uplinks share the same general geometry, even though the purpose of the link is completely different. Each has a transmitter on or near the Earth, each has a receiver in the satellite, and each must pass through most or all of the atmosphere. Losses include the space loss, atmospheric loss, rain loss, and antenna misalignment losses at the transmitter and the receiver.

The uplink equation is:

$$P_R = P_T + G_T - \text{Link Losses} + G_R$$

where P_R is the power received by the satellite receiver, P_T is the ground-based transmitter output power, G_T is the boresight gain of the ground-based transmitting antenna, *Link Losses* are those listed in Table 6.1, and G_R is the boresight gain of the satellite payload receiving antenna.

The losses described in Table 6.1 are defined in Chapter 5, which also includes the associated formulas.

At this point, it is important to note that the detailed link budgets of actual satellite systems often include other sources of loss, such as radome attenuation, that are left out of this discussion for simplicity.

Table 6.1
Uplink Losses

Loss	Description
Transmit antenna misalignment	Reduction from boresight gain at the offset angle from the direction to the satellite
Receiving antenna misalignment	Reduction from boresight gain at the offset angle from the direction to the ground transmitter
Space loss	LOS loss assuming that both the transmit and receive antennas are isotropic
Atmospheric loss	Atmospheric attenuation at the elevation angle through the atmosphere from the ground transmitter
Rain loss	Rain attenuation over the part of the link that passes through the rain

6.2.1 Command Links

Command links control the functions of the satellite such as its orientation, or they control the functions of the payload or payloads carried by the satellite. Different links can be used for the satellite and each payload, but a common link can be used to control some or all. The link signal will include addresses for each subsystem controlled and for each function of each subsystem.

Figure 6.7 shows the bit structure of a typical command link. Since a transmitted link must use a serial format, synchronization is required to allow the received signal processing to determine the function of each bit sent. This figure shows a very simple message structure, but real command messages can include dozens of functions. The synchronization can be very simple (e.g., one wide pulse). It can also be much more complex involving recognizable sequences of bit patterns. The synchronization identifies the beginning of a command frame. Then the other information sent can be identified by its position in that frame. Each frame can include multiple subframes.

Once synchronization is achieved, the commands for each function will include address bits and data bits. An address identifies the subsystem being commanded or the specific function being controlled. For example, if the ground station commands the satellite to a different orientation in order to point a satellite antenna in a desired direction, the navigation subsystem would be identified in the address and the desired azimuth and elevation would be specified in the data.

The payload functions can also be commanded. For example, if a payload includes a receiver, the ground station can tune to a desired frequency. In that case, the address bits would indicate which receiver is to be commanded and the data bits would specify the frequency to which it is to be tuned. In most cases, there will be many different satellite and payload functions to be commanded, so the data rate and message structure must be designed to ensure that each function is performed at the required speed to avoid unacceptable latency.

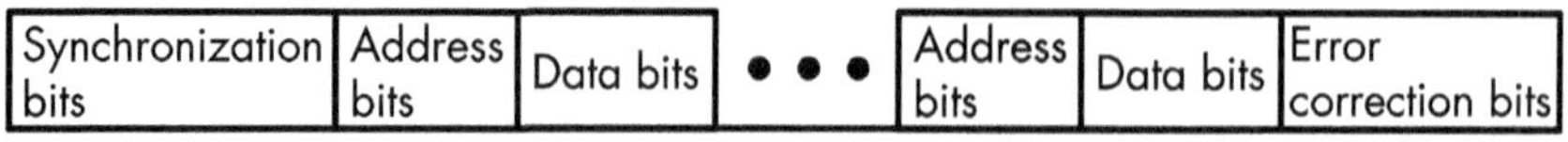

| Synchronization bits | Address bits | Data bits | • • • | Address bits | Data bits | Error correction bits |

Figure 6.7 The signal carried by a digital link includes synchronization, address, data, and error correction bits.

6.2 UPLINKS

It is important to assure that commands are received correctly by the satellite. Although the bandwidth of command signals is typically fairly low, the bandwidth of command signals can be significantly increased to accommodate error detection and correction approaches. In Chapter 7, the various types of error detection and correction approaches are described.

6.2.2 Intercept Links

If a satellite-borne system is to intercept hostile signals, there is a link defined from the hostile transmitter to the satellite payload receiver as shown in Figure 6.8. The hostile transmitter can be associated with a radar, a communication system, a broadcast station, or a data link. It can have any type of modulation. It can be on the ground or in an aircraft. In later chapters, we will deal with the performance that a satellite intercept system can achieve based on the satellite orbit, the nature of the transmitter, and the payload capability. However, right now, we will deal with the link in a general way.

The power received by the satellite payload receiver is given by:

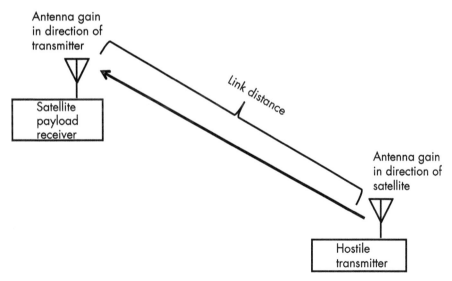

Figure 6.8 An intercept link takes signals from a hostile transmitter to a receiver in a satellite payload.

$$P_R = P_T + G_T - \text{Link Losses} + G_R$$

where P_R is the power received by the satellite payload receiver, P_T is the hostile transmitter output power, G_T is the boresight gain of the hostile transmitter, *Link Losses* are those listed in Table 6.1, and G_R is the boresight gain of the satellite payload receiving antenna.

Note that the intercepted transmitter antenna may have two different gain values. One is that to its desired receiver and a different gain to the intercept receiver in our satellite. If the target transmitter has an extremely wide angle antenna (like a dipole or a whip), these two gains may be equal. One geometric consideration is that both of these types of antennas have a null at 90° from the horizon if they are mounted vertically. This means that if the satellite is directly over the enemy transmitter, the satellite may not be able to receive the signal. Also, the signal polarization to the satellite may be different from that associated with transmission to its intended receiver. That may reduce the power of the signal received by the satellite. It is good practice to assume that the hostile transmitter has optimum antenna pointing and polarization to its intended receiver.

The location of the hostile transmitter and the satellite will determine the link distance. The antenna orientations will determine the effective transmit and receive antenna gains and the polarization loss.

Examples, including orbital considerations, are given in Chapter 9.

6.3 DOWNLINKS

Downlinks from the satellite to the ground can be to the satellite's control station, to other receivers at stations that require information that the satellite has gathered, or to hostile receivers that intercept the satellite downlink. They can also be to hostile receivers that are jammed from the satellite. Figure 6.9 shows the downlink. In any case, the downlink equation is

$$P_R = P_T + G_T - \text{Link Losses} + G_R$$

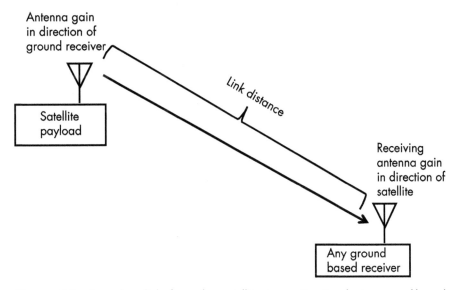

Figure 6.9 Any downlink from the satellite transmits signals to ground-based receivers.

where P_R is the power received by the ground-based receiver, P_T is the output power of the satellite's transmitter, G_T is the boresight gain of the satellite's antenna, *Link Losses* are those listed in Table 6.2, and G_R is the boresight gain of the ground-based receiving antenna.

6.3.1 Telemetry Link

A telemetry link passes information about the status and health of the satellite to its ground control station. This can include the satellite's orientation, the status of each of the satellite's systems, and the status and health of each of its payload functions. This is considered here as a separate link from any payload downlinks. When the satellite is first launched, this link allows a check of all satellite systems and received signal strength from command link transmissions. Among other functions, the telemetry link is important in the process of orienting a large satellite antenna toward the ground station. The telemetry link can be expected to have a wide coverage antenna.

6.3.2 Data Link

Data links carry information collected by a satellite payload to its ground control station. The format of these signals is basically that

Table 6.2
Downlink Losses

Loss	Description
Transmit antenna misalignment	Reduction from boresight gain at the offset angle from the direction to the ground-based receiver
Receiving antenna misalignment	Reduction from boresight gain at the offset angle from the direction to the satellite
Space loss	LOS loss assuming that both the transmit and receive antennas are isotropic
Atmospheric loss	Atmospheric attenuation at the elevation angle through the atmosphere from the ground-based receiver
Rain loss	Rain attenuation over the part of the link that passes through the rain

See Chapter 5 for the appropriate formulas.

shown for command links in Figure 6.7. However, the bandwidth of data links may be much wider than that of command links because of the amount of information output from payloads.

6.3.3 Links to Data Users

A satellite may collect data that is useful to multiple users. In this case, the satellite may be required to transmit that data directly to those users. These users may not be authorized to receive all of the data collected by the satellite; thus, it is typical to rebroadcast data from the ground control station back to the satellite in response to requests from users after the data has been edited. Then the edited data is transmitted to the authorized users who have requested it. These links are shown in Figure 6.10. The transmitting antenna from the satellite to those users will typically have a wide beamwidth, covering the Earth surface from horizon to horizon. The link equations for these links are as described for the uplinks and downlinks described earlier in this chapter.

6.3.4 Jamming Links

This section deals with jamming of ground communication links and radars from a satellite. Since the satellite is far from the ground, this is challenging, but it can be done in some circumstances. In this book, there is no guarantee that some kind of jamming is now done or practical. However, future technology may make it practical, so we consider it. Figure 6.11 shows a satellite to ground jamming link. Note

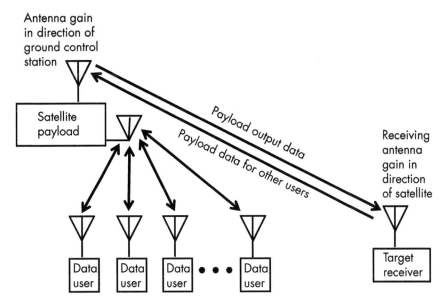

Figure 6.10 Information collected by a satellite can be processed for distribution to other users.

that the target receiver can be either on the Earth surface or in an aircraft flying above the Earth. Like intercept links, the received power into the target receiver is calculated by the same formula, but now the transmitted power is from the satellite jammer and the received power is at the target receiver

$$P_R = P_T + G_T - \text{Link Losses} + G_R$$

where P_R is the power received by the target receiver, P_T is the satellite jammer output power, G_T is the boresight gain of the jammer transmitter, *Link Losses* are those listed in Table 6.2, and G_R is the boresight gain of the target receiver antenna.

In this section, we will just give the appropriate formulas for the jamming links. Examples, including orbital considerations, are presented in Chapter 10.

6.3.4.1 Jamming a Ground Communication Link

The effectiveness of a jammer is determined by the amount of jammer-to-signal ratio (J/S) that it can cause in the target receiver. Figure 6.12

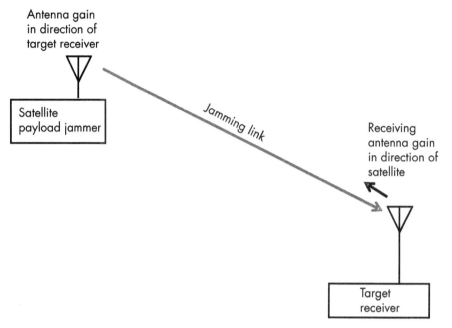

Figure 6.11 A satellite-borne jammer transmits a jamming signal to a receiver in a noncooperative target on or near the Earth.

shows the transmitter (hostile to us) that the target receiver is trying to receive. The J/S is the received power in the target receiver from the jammer divided by the received power in the target receiver from its desired signal (in this case, the hostile transmitter). Note that the target receiver's antenna may have two different gain values. One is that to its desired transmitter and a different gain to the jammer in our satellite. If the target receiver has an extremely wide angle coverage antenna (like a dipole or a whip), these two gains may be equal. If the target receiver antenna has a directional antenna, it is good practice to assume that it is directed toward the desired signal transmitter and has compatible polarization.

The J/S achieved is given by the formula

$$J/S = P_J + G_J - \text{Link Losses}_J + G_{RJ}$$
$$- \begin{bmatrix} P_T + G_T - \text{Atmospheric and Rain Losses in Atmosphere} \\ -\text{Applicable Link Losses} + G_R \end{bmatrix}$$

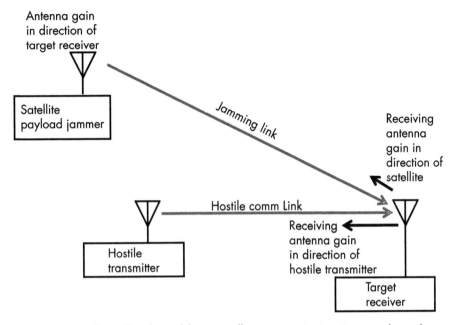

Figure 6.12 The J/S achieved by a satellite communication jammer depends on both the jamming link and the hostile communication link.

where J/S is the jamming-to-signal ratio in decibels, P_J is the satellite jammer output power (in dBm), G_J is the boresight gain of the jammer transmitter (in dBi), *Link Losses$_J$* are those listed in Table 6.2, for the jamming link, *Applicable Link Losses* are losses in the atmosphere affecting the jammed link (see Chapter 4), G_{RJ} is the gain of the target receiver antenna in the direction of the jammer (in dBi), P_T is the hostile transmitter output power (in dBm), G_T is the boresight gain of the hostile transmitter (in dBi), and G_R is the boresight gain of the target receiver antenna (in dBi).

As explained in Section 4.3, the target communication link losses are a function of the applicable link model. This can be LOS, two-ray, or knife edge diffraction, depending on the frequency, link distance, elevation, and surrounding geography.

6.3.4.2 Jamming Ground Radars

Figure 6.13 illustrates the links associated with jamming a ground-based radar from a satellite. Note that the ground radar (which we

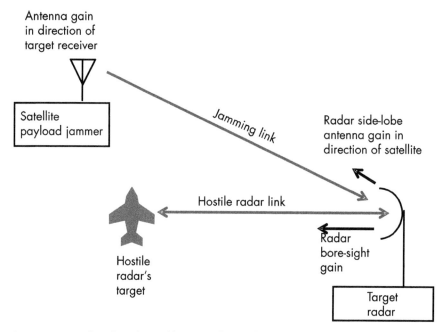

Figure 6.13 The J/S achieved by a satellite radar jammer depends on both the jamming link and the hostile radar link.

call the target radar) is tracking a target at some range and is receiving a skin return from that target. The boresight of the radar antenna is aimed at that target. The satellite's jamming signal is received by the radar's receiver, but since the radar is not aimed at the satellite, the radar antenna will receive the jamming signal in a sidelobe that will have significantly lower gain than the boresight gain. The effectiveness of the jammer is described in terms of the J/S that it can achieve. The J/S is the power ratio of the jamming signal received by the radar to the skin return power received by the radar. Note that, although radar signals can propagate using other propagation models, most have LOS propagation. Therefore, we will use that in this discussion.

We are assuming that the satellite jammer antenna is aimed at the target radar and that the target radar antenna is aimed at its target.

The J/S achieved is given by the following formula, which ignores rain and atmospheric loss both in the jamming and radar links:

$$J/S = ERP_J - ERP_S + 71 + 40\log R_T - 20\log R_J + G_S - G_M - 10\log RCS$$

where J/S is the jamming-to-signal ratio achieved (in decibels), ERP_J is the satellite jammer effective radiated power (in dBm), R_T is the range between the target radar and its target (in kilometers), ERP_s is the target radar's effective radiated power (in dBm), G_S is the sidelobe gain of the target radar in the direction of the satellite (in dBi), G_M is the main beam boresight gain of the target radar (in dBi), and RCS is the radar cross-section of the target radar's target in m².

Because the atmospheric and rain losses from the satellite can be significant and the satellite jamming antenna may be misaligned, the following adjustment to the J/S is presented. This J/S change value is added to the formula given above.

$\Delta J/S$ = Target Radar Atmospheric Loss (in dB)
+Target Radar Rain Loss (in dB)
− Satellite Atmospheric Loss (in dB) − Satellite Rain Loss (in dB)
− Satellite Jamming Antenna Misalignment Loss (in dB)

Each of these losses is defined and its formulas are given in Chapters 4 and 5.

6.4 HOSTILE LINKS

This chapter has described links that are associated with the operation of the satellite and its payloads. There are also links that are hostile. For example, if a hostile receiver is employed to intercept any of the satellite's downlinks, it would have a link as shown in Figure 6.14. Any of the uplinks to the satellite could also be intercepted. However, since the transmitters associated with uplinks are on the ground, the jammers must be located on airborne platforms unless they are very close to the jammed ground receiving sites. The links to these intercept receivers are shown in Figure 6.15.

Uplink and downlink jammers can also be used against satellites, and each involves a link. An uplink jamming link diagram

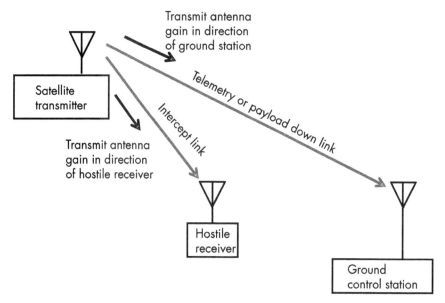

Figure 6.14 A hostile receiver could intercept the satellite downlink.

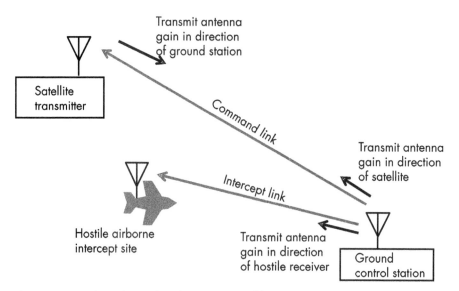

Figure 6.15 An airborne hostile receiver could intercept a satellite or payload uplink.

is shown in Figure 6.16 and a downlink jamming link diagram is shown in Figure 6.17.

Detailed coverage of these links is included in Chapter 7 as part of the link vulnerability discussion.

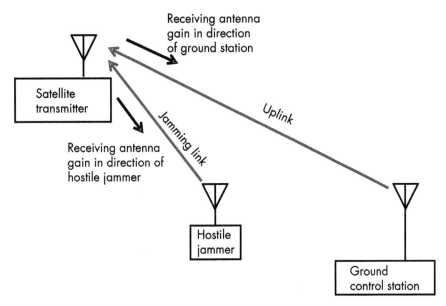

Figure 6.16 A hostile ground-based jammer could jam any satellite uplink.

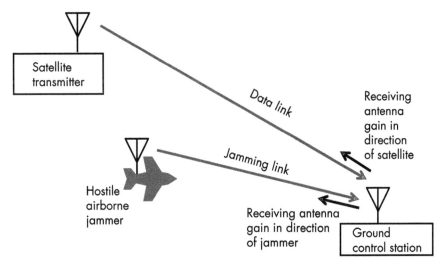

Figure 6.17 An airborne hostile jammer could jam any satellite downlink.

7

LINK VULNERABILITY TO EW

In this chapter, we will discuss the important subject of the vulnerability of satellites to threats from hostile electromagnetic attacks.

7.1 SATELLITE VULNERABILITY

Satellites are far from the Earth, but they present excellent LOS from a large part of the Earth's surface. Therefore, they are highly susceptible to three types of hostile activity. The signals from the satellite can be intercepted, and strong hostile transmissions can be jamming signals, interfering with uplink or downlink signals to prevent the signals from being properly received. They can also be spoofing signals that cause the satellite to interpret them as functional commands that are harmful.

We will start by determining the vulnerability of satellite links to intercept, spoofing, and jamming as though there was nothing that could be done to reduce that vulnerability. Then we will talk about the electronic protection measures that can reduce the vulnerability to both.

Intercept is illustrated in Figure 7.1 and spoofing is illustrated in Figure 7.2.

Successful intercept gives a hostile receiver a high enough quality signal that important information can be recovered. The received signals are intended for the receiver in the satellite's ground control station or some other authorized receiver. There is a separate link to

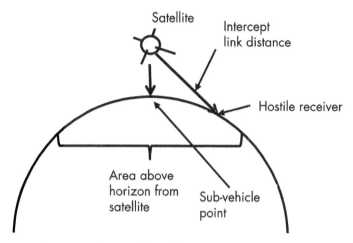

Figure 7.1 Intercept. A ground-based jammer operating against a satellite uplink transmits to the link receiver in the satellite. Both the ground station and the jammer must be above the horizon from the satellite.

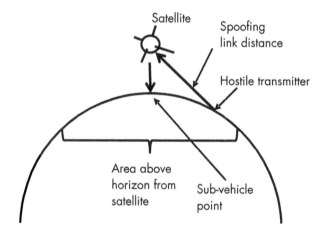

Figure 7.2 Spoofing. A ground-based jammer operating against a satellite uplink transmits to the link receiver in the satellite. Both the ground station and the jammer must be above the horizon from the satellite.

any hostile receiver. Successful spoofing places a strong enough signal into a satellite link receiver to cause the satellite or its payload to accept it as a valid command. Command spoofing could cause a satellite to perform a maneuver that ends its mission or could put the payload in an unusable state.

Uplink jamming is illustrated in Figure 7.3 and downlink jamming is illustrated in Figure 7.4. In both cases, the jammer transmits to a link receiver.

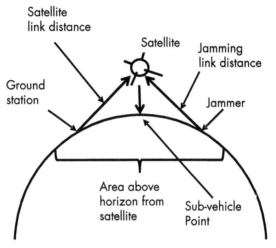

Figure 7.3 A ground-based jammer operating against a satellite uplink transmits to the link receiver in the satellite. Both the ground station and the jammer must be above the horizon from the satellite.

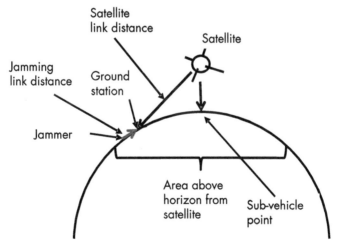

Figure 7.4 A jammer operating against a satellite downlink transmits to the link receiver in the satellite's ground station. The jammer can be on the Earth's surface or anywhere else that is above the horizon from the ground station.

Successful uplink jamming could prevent the proper function of the satellite or the payload, for example, preventing changes in satellite orientation or selection of a payload function.

Successful downlink jamming could prevent the ground station from knowing about a condition at the satellite that must be corrected. It could also prevent the transmission of payload data to the ground station.

7.1.1 Space-Related Link Losses

An attack on a satellite link may involve a single link, or it may involve multiple links. Each of these links is subject to transmission losses to include LOS loss, atmospheric loss, rain loss, and antenna losses from misalignment or polarization mismatch.

7.1.2 Intercept

An intercept link is separate from the intended command and data links. It goes from the satellite's link transmitter (on the satellite or in the control station) to a hostile receiver. The quality of the intercept is judged by the signal-to-noise ratio achieved in the hostile receiver.

7.1.3 Spoofing

A spoofing link goes from a hostile transmitter to a satellite link receiver. This receiver is typically on the satellite; the spoofing signal's purpose is to cause it to function improperly, but can be in the ground station if the purpose of the spoofing is to falsely imitate valid data.

7.1.4 Jamming

Jamming of any satellite link is communication jamming. Jamming effectiveness is normally defined in terms of the jamming-to-signal ratio (J/S) that it causes. Analysis of jamming requires that two links be considered. One is the link from an intended transmitter to an intended receiver. The second is from a jamming transmitter to that same receiver. The effectiveness of communication jamming is determined by whether or not the intended signal is properly received in the targeted receiver. In satellite link jamming, this is most often defined in terms of the J/S that is achieved. If noise jamming is employed, the

J/S can be as low at 0 dB, but there are circumstances in which much more J/S is required.

The *J/S* for communication jamming is calculated from the following formula:

$$J/S = ERP_J - ERP_S - LOSS_J + LOSS_S + G_{RJ} - G_R$$

where *J/S* is the jamming-to-signal ratio in decibels, ERP_J is the effective radiated power (ERP) of the jamming transmitter toward the target receiver in dBm, ERP_S is the ERP of the desired signal toward the receiver in dBm, $LOSS_J$ is the transmission loss from the jammer to the target receiver in decibels, $LOSS_S$ is the transmission loss from the transmitter to the target receiver in decibels, G_{RJ} is the gain of the receiving antenna in the direction of the jammer in decibels, and G_R is the gain of the receiving antenna toward the transmitter in decibels.

If the target receiver has a nondirectional antenna, the last two terms in this equation cancel each other out.

7.1.5 Problems Worked in This Chapter

In this chapter, there are five different scenarios considered. They are:

- Intercept of a downlink;
- Intercept of an uplink;
- Jamming of a downlink;
- Jamming of an uplink;
- Electronic protection of any satellite link.

All of the scenarios must consider whether the satellite is above the horizon to the related ground element. Note that there is a large table of the horizon distances versus satellite periods in Table 3.3 with the underlying calculations. This table will be very convenient when you use the presented information to work real-life problems. In this table, *p* (min) is the period of the satellite in minutes, *rng* (km) is the link propagation distance between the satellite and the Earth surface transmitter or receiver, and *dist* (km) is the Earth surface distance between the SVP and the Earth surface transmitter or receiver.

7.2 DOWNLINK INTERCEPT

For this discussion, we will assume that the vulnerable satellite is in a circular orbit 300 km above the Earth. Its SVP is at 100° East longitude and its latitude is 40° North. The intercepting site is on the Earth at 102° East longitude and 42° North latitude.

The satellite downlink has a 100-W (50-dBm) transmitter at 2 GHz. Note that this is just a calculation number; we don't care if any satellite has ever operated with this transmitter.

We will start by assuming that both the satellite downlink and the intercepting station have isotropic antennas with 0-dB gain. Later, we will change these to directional antennas.

There are several important values that we need to calculate:

- What is the range from the satellite to the intercept site?
- What is the elevation of the intercept site from the satellite?
- What is the elevation of the satellite from the intercept site?

7.2.1 Geocentric Angle from the Satellite to the Intercept Site

Figure 7.5 shows a spherical triangle on the Earth's surface formed by the North Pole, the satellite SVP, and the hostile receiving site. All

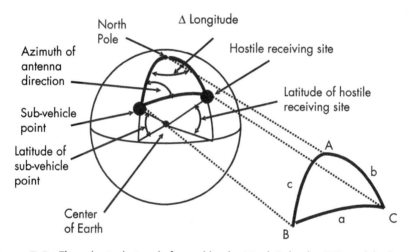

Figure 7.5 The spherical triangle formed by the North Pole, the SVP, and the hostile receiving site allows calculation of the satellite antenna azimuth to the hostile site and the geocentric angle between the satellite and the hostile site.

7.2 DOWNLINK INTERCEPT

three sides are pieces of great circle paths on the Earth's surface. In a spherical triangle on the Earth, the length of a side is stated as the geocentric angle between its end points. The dimension of an angle is the geocentric angle between the two planes on which the two adjacent sides are defined. Therefore, both the sides and the angles are expressed as angles.

Side a is the geocentric angle between the satellite SVP and the intercept site. Side b is 90° minus the latitude of the SVP. Side c is 90° minus the latitude of the intercept site. Angle A is the difference in longitude between the subvehicle location and the intercept site.

In Figure 7.5:
Side $a = 90° - 40° = 50°$
Side $b = 90° - 42° = 48°$
Angle A = 2°
The spherical law of cosines for sides is:

$$\cos a = (\cos b)(\cos c) + (\sin b)(\sin c)(\cos A)$$
$$= (\cos 50°)(\cos 48°) + (\sin 50°)(\sin 48°)(\cos 2°)$$
$$= (0.643)(0.669) + (0.766)(0.743)(0.999)$$
$$= 0.430 + 0.569 = 0.999$$

So side a = 2.56°.

7.2.2 Range from the Satellite to the Intercept Site

Now consider Figure 7.6. This is a plane triangle formed by the satellite, the center of the Earth, and the intercept site. Side d is the range from the satellite to the intercept site. Side e is the radius of the Earth (R_E) plus the elevation of the satellite (H). Side f is the radius of the Earth. Angle D is the same angle that we calculated as a in the spherical triangle of Figure 7.5.

Angle D is 2.56°. This is the same as side a in the spherical triangle of Figure 7.5.
Side e is 6,671 km.
Side f is 6,371 km.
The law of cosines for sides in plane triangles is:

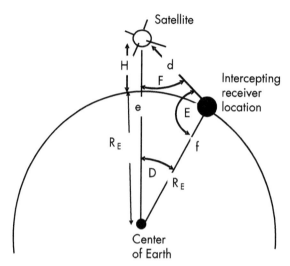

Figure 7.6 The propagation distance between a transmitting satellite and an intercept receiver on the Earth's surface can be calculated from the plane triangle formed by the satellite location, the receiver location, and the center of the Earth.

$$d^2 = e^2 + f^2 - 2ef \cos D$$
$$= (6{,}671 \text{ km})^2 + (6{,}371 \text{ km})^2 - 2(6{,}671)(6371) \cos 2.56°$$
$$= 44{,}502{,}241 + 40{,}589{,}641 - 84{,}916{,}880 \text{ km}^2$$
$$= 175{,}002 \text{ km}^2$$
$$d = sqrt(175{,}002 \text{ km}^2)$$
$$= 418 \text{ km}$$

Continuing with Figure 7.6, we want to calculate the elevation from the satellite to the intercept site and the elevation angle from the intercept site to the satellite. We know side d and angle D, so we can use the law of sines for plane triangles.

$\sin D / d = \sin E / e = \sin F / f$
$\sin E = e \sin D / d = 6671 \sin(2.56°) / 418 = 6671(0.0493) / 418 = .713$
$\arcsin(.787) = 45.5°$

7.2 DOWNLINK INTERCEPT

but angle E is greater than 90°, so angle E (the elevation of the satellite above the center of the Earth) = 180° − 45.5° = 134.5°. The elevation of the satellite above the horizon viewed from the intercept site is 45.5°.

$$\sin F / 6{,}371 = \sin 2.56° / 418 = 0.0447 / 418 = 0.000{,}107$$
$$\sin F = (0.000{,}107)(6{,}371) = .681$$

Angle F (the elevation of the intercept site above the center of the Earth) = 42.9°

The Earth surface distance from the SVP to the intercept site is given by the formula:

$$(\text{Angle } D / 360°)(2\pi \times \text{Radius of the Earth})$$
$$= (2.56° / 360°)(2\pi \times 6371 \text{ km}) = 285 \text{ km}$$

We will use two of these three numbers (link distance and elevation of the satellite above the horizon from the intercept site) in determining the intercept link performance.

7.2.3 Is the Satellite Above the Horizon from the Intercept Site?

Now consider Figure 7.7. It shows a plane right triangle made by the satellite, the horizon point, and the center of the Earth.

Angle J is 90°, and it is the angle from the center of the Earth to the satellite as seen from the horizon point.

Angle K is the geocentric angle from the satellite to the horizon.

Angle M is the angle from the center of the Earth to the horizon as seen from the satellite.

Side j is the distance from the center of the Earth to the satellite ($R_E + H$).

Side k is the direct distance from the satellite to the horizon.

Side m is the distance from the center of the Earth to the horizon point (R_E).

Since this is a plane right triangle, side j^2 = side k^2 + side m^2.

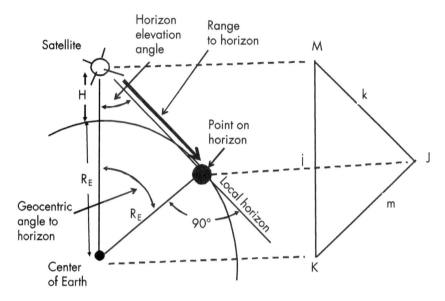

Figure 7.7 The range to the horizon from a satellite can be determined from the plane triangle defined by the satellite, a point on the horizon, and the center of the Earth.

$$\text{Side } k = \text{sqrt}(\text{side } j^2 - \text{side } m^2)$$
$$= \text{sqrt}(6671^2 - 6371^2) = \text{sqrt}(44,502,241 - 40,589,641)$$
$$= \text{sqrt}(3,912,600)$$
$$= 1978 \text{ km}$$

This is the signal path to the horizon, so the intercept site is well within the distance to the horizon.

Angle K can be found from the formula (based on the law of cosines for sides for plane triangles):

$$\cos K = (j^2 + m^2 - k^2)/2jm$$
$$= (6671^2 + 6371^2 - 1978^2)/2 \times 6671 \times 6371$$
$$= (44,502,241 + 40,589,641 - 3,912,484)/85,001,882$$
$$= 81,179,398/85,001,882$$
$$= 0.955$$

Angle $K = a \cos (0.955) = 17.2°$

The Earth surface distance from the SVP to the horizon is given by the formula:

$$(\text{Angle } K / 360°)(2\pi \times \text{Radius of the Earth})$$
$$= (17.2° / 360°)(2\pi \times 6{,}371 \text{ km}) = 1{,}913 \text{ km}$$

The intercept site is well within the horizon distance from the SVP.

7.2.4 How Strong Is the Downlink Signal at the Intercept Site

The received signal, ignoring atmospheric, rain, and polarization losses and several more important considerations, is given by the formula:

$$P_R = ERP - (32.4 + 20\log F + 20\log d)$$

where P_R is the received signal power at the intercept site in dBm, *ERP* is the ERP from the satellite in dBm, F is the link transmission frequency in megahertz, and d is the link distance in kilometers.

$$P_R = 50 - 32.4 - 20\log(2{,}000) - 20\log(418)$$
$$= 100 - 32.4 - 66 - 52.4 = -100.8 \text{ dBm}$$

Since at this point we have assumed isotropic (i.e., 0-dB) gain, this may be enough signal, depending on the type and bandwidth of the receiver.

7.2.5 Intercept with Directional Antennas

Now we will significantly complicate this intercept problem by adding directional antennas, several more loss elements, signal modulation, and receiver sensitivity.

We will use the same satellite: the vulnerable satellite is in a circular orbit 300 km above the Earth. Its SVP is at 100° East longitude and its latitude is 40° North. The satellite's ground control station is on the Earth at 103° East longitude, 44° North latitude. There is a hostile intercept site on the Earth at 102° East longitude and 45° North latitude. The satellite downlink has a 100-W (50-dBm) transmitter at 2 GHz. Note that this is just a calculation number; it is not representative of any specific satellite or intercept site.

There are several diagrams in this discussion dealing with relative positions of the satellite and ground locations. Please be aware that the angles shown in these diagrams are not drawn to scale; they are spread out to allow labeling.

We will give both the satellite downlink and the hostile ground intercept station directional antennas. The transmitting antenna on the satellite is a 3-m parabolic dish. Both the satellite ground station and the intercepting ground station have 2-m parabolic dishes. First, we want to know the gain and 3-dB beamwidth of each of these antennas.

The antenna boresight gain of each antenna can be determined from the formula:

$$G = -42.2 + 20\log D + 20\log F$$

where G is the boresight gain in dBi, D is the diameter of the antenna in meters, and F is the operating frequency in megahertz.

For the satellite antenna,

$$G = -42.2 + 20\log(3) + 20\log(2000)$$
$$G = -42.2 + 9.5 + 66 = 33.3 \text{ dBi}$$

For the ground station antenna and also for the intercept antenna,

$$G = -42.2 + 20\log(3) + 20\log(2000)$$
$$G = -42.2 + 6 + 66 = 29.8 \text{ dBi}$$

The 3-dB beamwidth of each antenna can be found from the formula:

$$\alpha = \text{Antilog}\left[(86.8 - 20\log D - 20\log F)/20\right]$$

where α is the 3-dB beamwidth in degrees, D is the diameter of the antenna in meters, and F is the operating frequency in megahertz.

For the satellite antenna,

$$\alpha = \text{Antilog}\big[(86.8 - 20\log(3) - 20\log(2000))/20\big]$$
$$\alpha = \text{Antilog}\big[(86.8 - 9.5 - 66)/20\big] = 3.7°$$

For the intercept site and the ground control station antennas,

$$\alpha = \text{Antilog}(86.8 - 6 - 66)/20] = 5.4°$$

7.2.6 Angles Relative to the Ground Control Station

Now we will determine the look angles from the satellite to its ground control station. Figure 7.8 shows a spherical triangle formed by the North Pole, the SVP, and the satellite's ground control station. The dimensions of the triangle are:

- Angle A = the difference in longitude between the satellite and the ground station = 3°.
- Angle B is the azimuth from the satellite to the ground station.
- Side c = 90° minus the latitude of the SVP = 50°.
- Side b = 90° minus the latitude of the ground station = 46°.

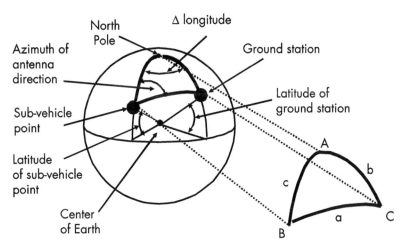

Figure 7.8 The spherical triangle formed by the North Pole, the SVP, and the ground station allows calculation of the satellite antenna azimuth to the ground station and the geocentric angle between the satellite and the ground control station.

The spherical law of cosines for sides can be written as:

$$\cos a = \cos b \times \cos c + \sin b \times \sin c \times \cos A$$
$$= \cos(46°) \times \cos(50°) + \sin(46°) \times \sin(50°) \times \cos(3°)$$
$$= (.695)(.643) + (.719)(.766)(.999) = .446 + .550 + .996$$

Side a = arc cos(0.997) = 5.13°

This is the geocentric angle between the SVP and the ground station.

Note that the sine of this angle is 0.089.

The spherical law of cosines for sides can be reorganized to read as:

$$\cos B = (\cos b - \cos a \times \cos c)/(\sin a \times \sin c)$$
$$= [(.695) - (.996)(.643)]/(.089)(.766) = .055/.068 = .809$$

Angle B is arc cos(0.809) = 36.02°.

This is the azimuth to which the satellite antenna must be oriented to aim at the ground station.

Now we will determine the elevation of the ground control station from the satellite using Figure 7.9. This is a plane triangle with angles at the satellite, the ground control station, and the center of the Earth.

- Angle F is the same as angle a in Figure 7.8 = 5.13°.
- Side e is the radius of the Earth + the height of the satellite = 6,671 km.
- Side d is the radius of the Earth = 6,371 km.

The law of cosines for sides for plane triangles is:

$$\text{Side } f^2 = e^2 + d^2 - 2ed \cos F$$
$$= 6671^2 + 6371^2 - 2(6671)(6371)(\cos 5.13°)$$
$$= 44,502,241 + 40,589,641 - 84,661,397$$
$$= 446,732 \ km^2$$

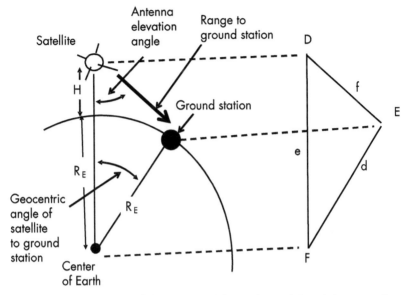

Figure 7.9 The elevation of the antenna (above the nadir) and the range from the satellite to the ground station can be calculated from this plane triangle.

Side f is the square root of this number = 668 km. This is the link distance from the satellite to its ground control station.

Now we can use the law of sines for plane triangles to find angle D, the elevation of the ground station from the satellite. Remember that the elevation is the angle up from the center of the Earth.

$$\sin D = (d \times \sin F) / f = [6371 \; km \times \sin(5.13°)] / 668 \; km = .853$$

Angle D = arcsin (.0756) = 58.3°.

This is the elevation of the ground station as seen from the satellite.

7.2.7 Angles Relative to the Hostile Intercept Site

Now we will determine the look angles from the satellite to the intercept site. Figure 7.10 shows a spherical triangle formed by the North Pole, the SVP, and the intercept site. The dimensions of the triangle are:

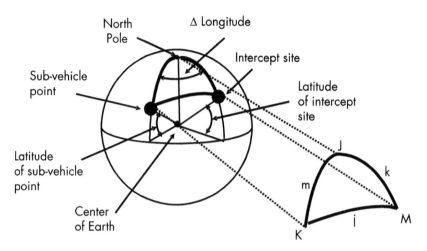

Figure 7.10 A spherical triangle between the North Pole, the SVP, and the intercept site allows calculation of the offset angle of the intercept site antenna from the satellite downlink antenna boresight.

- Angle J = the difference in longitude between the satellite and the intercept site = 2°.
- Angle K is the azimuth from the satellite to the intercept site station.
- Side m = 90° minus the latitude of the SVP = 50°.
- Side k = 90° minus the latitude of the intercept = 45°.

The spherical law of cosines for sides can be written as:

$$\cos j = \cos m \cos k + \sin m \times \sin k \times \cos J$$
$$= \cos(50°) \times \cos(45°) + \sin(50°) \times \sin(45°) \times \cos(2°)$$
$$= (.642)(.707) + (.766)(.707)(.999) = .454 + .541 = .995$$

Side j = arc cos(0.995) = 5.73°

This is the geocentric angle between the SVP and the intercept site. Note that the sine of this angle is 0.099.

The spherical law of cosines for sides can be reorganized to read as:

$$\cos K = (\cos k - \cos j \times \cos m)/(\sin j \times \sin m)$$
$$= [(.707) - (.995)(.642)]/(.099)(.766) = .069/.0758 = .910$$

7.2 DOWNLINK INTERCEPT

Angle K is arc cos(0.910) = 24.5°.

This is the azimuth to which the satellite antenna must be oriented to aim at the intercept site.

Now we will determine the elevation of the intercept site from the satellite using Figure 7.11. This is a plane triangle with angles at the satellite, the intercept site, and the center of the Earth.

- Angle P is the same as angle g in Figure 7.10 = 5.73°.
- Side r is the radius of the Earth + the height of the satellite = 6,671 km.
- Side q is the radius of the Earth = 6,371 km.

The law of cosines for sides for plane triangles is:

$$\text{Side } p^2 = r^2 + q^2 - 2rq\cos P$$
$$= 6{,}671^2 + 6{,}371^2 - 2(6{,}671)(6{,}371)(\cos 5.73°)$$
$$= 44{,}502{,}241 + 40{,}589{,}641 - 84{,}577{,}164$$
$$= 514{,}718 \text{ km}^2$$

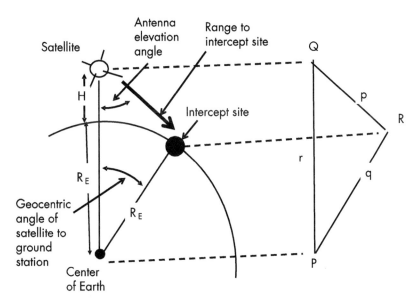

Figure 7.11 The elevation of the antenna (above the nadir) and the range from the satellite to the ground station can be calculated from this plane triangle.

Side f is the square root of this number = 717 km. This is the link distance from the satellite to the intercept site.

Now we can use the law of sines for plane triangles to find angle Q, the elevation of the intercept site from the satellite. Remember that the elevation is the angle up from the center of the Earth.

$$\sin Q = (q \times \sin P) / p = [6371 \ km \times \sin(5.73°)] / 717 \ km = .887$$

Angle Q = arc sin(0.887) = 62.5°

This is the elevation of the ground station as seen from the satellite.

Let's collect some of these numbers for the two downlinks.

- Downlink receiver direction from the satellite:
  - Azimuth = 36.0°;
  - Elevation = 58.3°.
- Intercept receiver direction from the satellite:
  - Azimuth = 24.5°;
  - Elevation = 62.5°.

The difference in azimuth between the downlink antenna boresight and the intercept site is $\Delta AZ = 24.5° - 36.0° = 11.5°$

The difference in elevation between the downlink antenna boresight and the intercept site is $\Delta EL = 62.5° - 58.3° = 4.2°$.

The offset angle of the direction to the intercept site from the downlink boresight is determined from the spherical right triangle in Figure 7.12.

From Napier's rules explained in Section 2.3, the cosine of side cc is the product of the cosines of sides aa and bb.

Side aa is $\Delta az = 11.5°$; cos aa = 0.980.
Side bb is $\Delta el = 4.2°$; cos bb = 0.997.
So cos cc = (0.980)(0.997) = 0.977, which makes side cc = 12.3°.

This is much more than the antenna beamwidth, so we can use an average sidelobe isolation of 20 dB to estimate the gain of the downlink antenna in the direction of the intercept antenna. Thus the gain of the downlink antenna in the direction of the intercept receiver is 20 dB below its boresight gain = 33.3 dB – 20 dB = 13.3 dB.

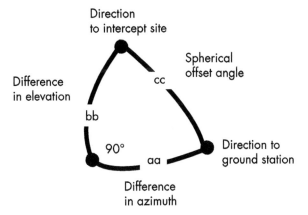

Figure 7.12 This right spherical triangle is on the surface of a unit sphere centered on the satellite downlink antenna. It is formed by the intercept site and the ground station, and can be defined by the difference in azimuth and difference in elevation, both as seen from the satellite.

7.2.8 Received Signal at the Intercept Site

Because of the high elevation angles, atmospheric and rain attenuations are negligible, so the loss is only LOS loss.

$$P_R = P_T + G_T (\text{Toward Receiver}) - 32.44 - 20\log d - 20\log F + G_R$$

where P_R is the power received by the receiver at the intercept site in dBm, P_T is the downlink transmitter power in dBm = 100W (i.e., 50 dBm), G_T is the transmit antenna gain toward the receiver = 13.3 dBi, d is the link distance between the satellite and the intercept site in kilometers = 717 km, F is the link frequency = 2 GHz (i.e., 2,000 MHz), and G_R is the receiving antenna gain at the intercept site (in dBi) = 29.8 dBi.

$$\begin{aligned} P_R &= P_T + G_T (\text{Toward Receiver}) - 32.4 - 20\log d - 20\log F + G_R \\ &= 50 \text{ dBm} + 13.3 \text{ dBi} - 32.4 - 20\log(717) - 20\log(2{,}000) + 29.8 \text{ dB} \\ &= 50 + 13.3 - 32.4 - 57.1 - 66 + 29.8 = -62.4 \text{ dBm} \end{aligned}$$

7.2.9 What Is the Quality of the Intercepted Downlink Signal?

The minimum detectable signal of the receiver is:

$$MDS = kTB + NF$$

where *MDS* is the minimum detectable signal in dBm, *kTB* is the thermal noise level in the receiver (which equals −114 dBm + 10 log(Bandwidth/1 MHz)), and *NF* is the receiver system noise figure in decibels.

If the bandwidth of the intercept receiver is 10 MHz and its noise figure is 5 dB, the receiver's MDS would be $kTB + NF = -104$ dBm + 5 dB = −99 dBm, and it would receive this signal with a signal-to-noise ratio of:

$$S/N = -62.4 \text{ dBm} - (-99 \text{ dBm}) = 36.6 \text{ dB}$$

This is a very high-quality intercept.

7.3 INTERCEPTING UPLINKS

Intercepting uplinks to satellites involves receiving signals from the ground station that are sent to the satellite. Unless the intercepting station is extremely close to the ground station, this can only be accomplished by looking down from a (perhaps unmanned) aircraft flying over the satellite's ground station as shown in Figure 7.13. Note that the aircraft must be above the local horizon from the ground station.

In order to determine the quality of the intercept, it will be necessary to know the distance from the aircraft to the ground station and the sidelobe isolation of the satellite link's receiving antenna. The ground station antenna is aimed at the satellite, so the intercepting aircraft will be in a sidelobe with reduced antenna gain. By making the assumption that the aircraft is not in the main beam of the satellite uplink antenna, we can just use its specified average sidelobe gain reduction rather than using precise antenna pattern data. In setting up this problem, we need to place the aircraft at a specified location. The latitude and longitude of the ground station and the latitude, longitude, and altitude (relative to the ground station) of the aircraft need to be input.

We start with Figure 7.14. The SVP of the aircraft, the ground station, and the North Pole form a spherical triangle.

- Side c is 90° minus the latitude of the aircraft SVP.

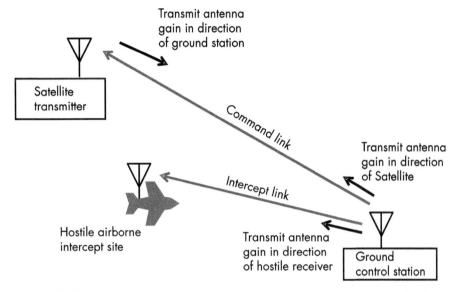

Figure 7.13 An airborne hostile receiver could intercept a satellite or payload uplink.

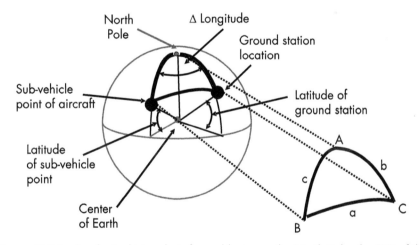

Figure 7.14 A spherical triangle is formed between the North Pole, the SVP of the aircraft, and the ground station.

- Side b is 90° minus the latitude of the ground station location.
- Angle A is the difference in longitude between the aircraft and the ground station.

Using the spherical law of cosines for sides, we can calculate the geocentric angle from the aircraft to the ground station as

$$\text{Side } a = \arccos\left[(\cos b)(\cos c) + (\sin b)(\sin c)(\cos A)\right]$$

Now consider Figure 7.15, which is a plane triangle formed by the aircraft, the ground station, and the center of the Earth.

- Side d is the radius of the Earth increased by the altitude of the ground station.
- Side e is the radius of the Earth increased by the altitude of the aircraft.
- Angle F is the same as side a from Figure 7.14.
- Side f is the range from the aircraft to the ground station.

The range is found using the planar law of cosines for sides:

$$\text{Side } f = \text{sqrt}\left[d^2 + e^2 - 2de\cos F\right]$$

Let's plug in some numbers:

- The ground station is at sea level at 45° North latitude and 100° East longitude.

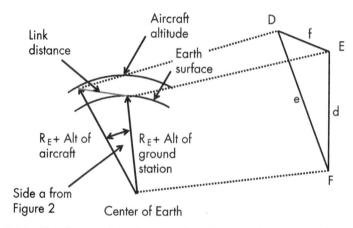

Figure 7.15 The distance from the aircraft to the ground station can be calculated from this plane triangle.

7.3 INTERCEPTING UPLINKS

- The aircraft is at 46° North latitude, 101° East longitude, and 3-km altitude.
- The uplink transmit frequency is 5 GHz.
- The uplink transmitter power is 100W (+ 50 dBm).
- The uplink ground antenna is a 3-m parabolic dish with circular polarization.
- The average sidelobe from the satellite's ground antenna is 20 dB below the boresight gain.
- The intercept payload antenna on the aircraft is a cavity-backed spiral with matched circular polarization and 3-dB gain.

The geocentric angle from the aircraft to the ground station is:

$$\arccos[(\cos 45°)(\cos 44°) + (\sin 45°)(\sin 44°)\cos(1°)]$$
$$= \arccos[(.7071)(.7193) + (.7071)(.6947)(.9998)] = \arccos[(.5086) + (.4911)]$$
$$= \arccos[.9998] = 1.403°$$

From the planar law of cosines for sides, the range is:

$$\text{sqrt}[(6371^2)(6374^2) - 2(6371)(6374)(.9997)]$$
$$= \text{sqrt}[(40589641) + (40627876) - (81193143)] = \text{sqrt}[24374] = 156 \text{ km}$$

This is the intercept link propagation range.

Since the satellite uplink frequency is 5 GHz and the transmitting antenna is a parabolic dish with 3-m diameter, its gain can be calculated from the formula:

$$\text{Gain} = -42.2 + 20\log(\text{Diameter in Meters}) + 20\log(\text{Frequency in MHz})$$
$$= -42.2 + 20\log(3) + 20\log(5000) = -42.2 + 9.5 + 74 = 41.3 \ dB$$

The aircraft will be in a sidelobe that is 20 dB weaker than the boresight gain, so the effective antenna gain to the intercept receiver is 21.3 dBi. Thus, with 50-dBm transmitter power, the ERP of the uplink toward the aircraft is 50 dBm + 21.3 dBi = 71.3 dBm.

The received power at the intercept receiver is:

$$P_R = ERP_T - 32.4 - 20\log d - 20\log F + G_R$$

where P_R the received power, ERP_T is the ERP of the uplink transmitter, d is the distance between the ground station and the intercept aircraft, F is the uplink transmitting frequency, and G_R is the gain of the intercept receiver's antenna.

Plugging in our numbers, the intercept receiver in the aircraft will receive:

$$P_R = 71.3 - 32.4 - 20\log(156) - 20\log(5,000) + 3$$
$$= 71.3 - 32.4 - 43.8 - 74 + 3 = -75.9 \text{ dBm}$$

This is a strong signal. Since the intercepting aircraft is above the horizon from the ground station, a successful intercept has taken place. This assumes that the intercept aircraft is allowed to fly over the satellite ground station without being shot down.

7.4 JAMMING DOWNLINKS

Jamming downlinks from satellites involves transmitting signals into the ground station. Unless the jamming transmitter is extremely close to the ground station, this can only be accomplished by looking down at the intercept site from an aircraft (perhaps unmanned) flying over the satellite's ground station as shown in Figure 7.16. Note that the aircraft must be above the local horizon from the ground station.

Since we have already done the hard work of calculating distances from the aircraft in Section 7.3, let's look at the jamming of downlinks by that same aircraft. For convenience, let's put the aircraft and ground stations in the same location and keep the aircraft and ground antennas the same and jam at the same frequency. Actually, the uplink and downlink need to be at slightly different frequencies, but we will use the same 5-GHz number for convenience. The jammer transmitter has 10-W (40-dBm) ERP. The distance from the jammer to the ground station is 156 km. The satellite is in a 1,000-km-high orbit and its SVP is 55° North latitude, 108° East longitude. The satellite's downlink ERP is 10W (40 dBm).

As always in this book, these values do not represent any specific satellite system. They are just chosen as reasonable values so we

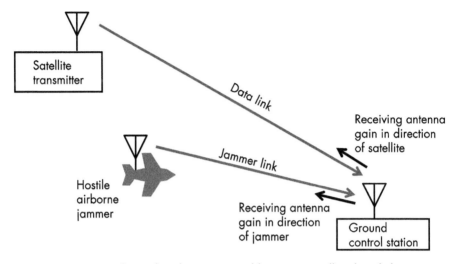

Figure 7.16 An airborne hostile jammer could jam any satellite downlink.

can look at the math. When you are performing design or analysis of a real-world satellite system, you can plug the real-world values into the equations we will discuss here.

7.4.1 The Satellite Downlink

We start with Figure 7.17. The SVP of the satellite, the ground station, and the North Pole form a spherical triangle.

- Side c is 90° minus the latitude of the aircraft SVP.
- Side b is 90° minus the latitude of the ground station location.
- Angle A is the difference in longitude between the aircraft and the ground station.

Using the spherical law of cosines for sides, we can calculate the geocentric angle from the aircraft to the ground station as

$$\text{Side } a = \arccos\left[(\cos b)(\cos c) + (\sin b)(\sin c)(\cos A)\right]$$

Now consider Figure 7.18, which is a plane triangle formed by the aircraft, the ground station, and the center of the Earth.

- Side d is the radius of the Earth increased by the altitude of the ground station.

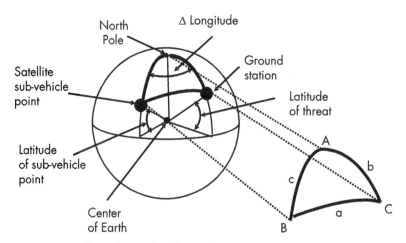

Figure 7.17 A spherical triangle is formed between the North Pole, the SVP, and the ground station location.

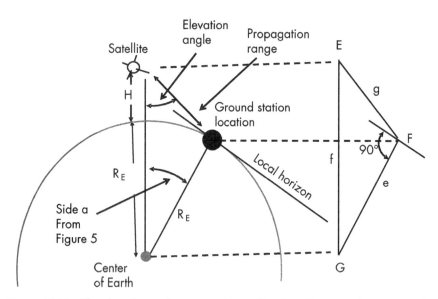

Figure 7.18 The elevation and range to the satellite from the ground station can be determined from the plane triangle defined by the satellite, the ground station, and the center of the Earth.

- Side e is the radius of the Earth increased by the altitude of the satellite.
- Angle F is the same as side a from Figure 7.17.

7.4 JAMMING DOWNLINKS

- Side f is the range from the satellite to the ground station.

The range is found using the planar law of cosines for sides:

Let's plug in some orbit numbers. The SVP is at longitude 108° East of the Greenwich meridian and 45° North latitude. The ground station location is at 100° East longitude, 55° North latitude. The satellite has a 1,000-m-high circular orbit and the radius of the Earth is 6,371 km.

For the spherical triangle in Figure 7.17:
Angle A = 8°.
Side b is 90° − 55° = 35°.
Side c is 90° − 45° = 45°.
We find side a from the spherical law of cosines for sides:

$$\text{Side } a = \arccos[(\cos 35°)(\cos 45°) + (\sin 35°)(\sin 45°)(\cos 8°)]$$
$$= \arccos[(.819)(.707) + (.574)(.707)(.990)]$$
$$= \arccos[.579 + .402] = \arccos[.981] = 11.2°$$

Now consider the plane triangle in Figure 7.18. Using the law of cosines for sides for plane triangles:

$$\text{Side } g = \text{sqrt}[e2 + f2 - 2ef \cos(G)]$$

Let's plug in some numbers:
Side f is 6,371 km + 1,000 km = 7,371 km.
Side e is 6,371 km.
Angle G is the same as side a in Figure 7.17 2 = 11.2°.

$$\text{Side } g = \text{SQRT}[(7,371)^2 + (6,371)^2 - 2(7,371)(6,371)\cos(11.2°)]$$
$$= \text{SQRT}[54,331,641 + 40,589,641 - 46,066,285]$$
$$= \text{SQRT}[48,854,997] = 6990 \text{ km}$$

This is the downlink distance.

Now, using the law of sines for plane triangles, the elevation angle from the satellite to the ground station is:

Angle $E = \arcsin\left[e(\sin G / g)\right]$

Angle $E = \arcsin[6371(\sin(11.2°))/6990]$

$\quad = \arcsin\left[(6371)(.1942)/6990\right] = \arcsin[.1770] = 10.2°$

The elevation of the satellite from the ground station is the angle above the local horizon. From Figure 7.18, the local horizon is shown to make a 90° angle to side e at the ground station location. This means that the elevation of the satellite from the ground station is angle F − 90°.

Since this is a plane triangle, the sum of the three internal angles is 180°, and we know that angle E is 10.2° and angle G is 11.2°. Thus, angle F is:

$$180° - 10.2° - 11.2° = 158.6°$$

The elevation above the local horizon is 158.6° − 90° = 68.6°. At this high angle, the atmospheric and rain losses are negligible at 5 GHz.

The LOS loss from the satellite to the ground station is

$$32.4 + 20\log(d) + 20\log(F) = 32.4 + 76.5 + 74 = 182.9 \text{ dB}$$

7.4.2 The Jammer Link

The jammer LOS loss is

$$32.4 + 20\log(d) + 20\log(F) = 32.4 + 20\log(156) + 20\log(5000)$$
$$= 32.4 + 43.9 + 74 = 150.3 \text{ dB}$$

7.4.3 The J/S Formula

The formula for communications jamming is

$$J/S = ERP_J - ERP_S - LOSS_J + LOSS_S + G_{SJ} - G_S$$

where J/S is the jamming-to-signal ratio in decibels, ERP_J is the ERP of the jammer in dBm, ERP_S is the ERP of the jammed signal in dBm,

$LOSS_J$ is the loss between the jammer and the jammed receiver in decibels, $LOSS_S$ is the loss between the desired signal transmitter and the jammed receiver in decibels, G_{SJ} is the gain of the jammed receiver's antenna in the direction of the jammer in decibels, and G_S is the gain of the jammed receiver's antenna in the direction of the desired signal transmitter in decibels.

Plugging in the values for this problem,

$$J/S = 40 \text{ dBm} - 40 \text{ dBm} - 150.3 \text{ dB} + 182.9 \text{ dB}$$
$$+ 21.3 \text{ dB} - 41.3 \text{ dB} = 12.6 \text{ dB}$$

This is effective jamming, again assuming that the jamming aircraft is allowed to fly over the satellite's control station without being shot down.

7.5 JAMMING SATELLITE UPLINKS

To jam the uplink to a satellite from the ground, the jamming signal must be received by the link receiver in the satellite. In order to determine the J/S, we must deal with the uplink from the ground station to the satellite and the link from the jammer to the satellite. We assume here that the jammer is on the Earth's surface. Here is a problem that illustrates the process.

The satellite uplink has a 10-W (40-dBm) transmitter at 5 GHz with a 2-m-diameter parabolic antenna. The jammer has a 1-kW (60-dBm) transmitter with a 4-m-diameter parabolic antenna. The uplink receiving antenna in the satellite is a 1-m parabolic antenna. All of these antennas have right-hand circular polarization, so all of the signals involved have matched polarization. The satellite is in a 300-km-high circular orbit. The satellite SVP is at 100° East longitude, 40° North latitude. The ground station is at 103° East longitude, 42° North latitude. The jamming site is at 102° East longitude, 45° North latitude. In this section, we will assume that the atmospheric and any rain loss are the same for the links from the ground station and the jammer, so they will not impact the calculations. To evaluate the effectiveness of the jamming, we must consider the jamming link and the satellite's uplink.

7.5.1 The Jamming Link

Figure 7.19 is a spherical triangle formed by the satellite SVP, the jammer location, and the North Pole. Side a is the geocentric angle between the satellite SVP and the jammer. Side b is 90° minus the latitude of the SVP. Side c is 90° minus the latitude of the jammer. Angle A is the difference in longitude between the subvehicle location and the jamming site.

Side $c = 90° - 40° = 50°$
Side $b = 90° - 45° = 45°$
Angle $A = 2°$

The spherical law of cosines for sides is:

$$\begin{aligned}\cos a &= (\cos b)(\cos c) + (\sin b)(\sin c)(\cos A) \\ &= (\cos 50°)(\cos 45°) + (\sin 50°)(\sin 45°)(\cos 2°) \\ &= (0.643)(0.707) + (0.766)(0.707)(0.999) \\ &= 0.454 + 0.541 = 0.995\end{aligned}$$

So side a = 5.73°.

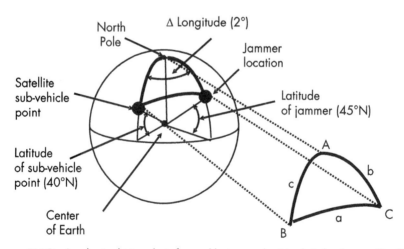

Figure 7.19 A spherical triangle is formed between the North Pole, the satellite SVP, and the transmitter location.

7.5 JAMMING SATELLITE UPLINKS

Angle B in Figure 7.19 is the azimuth from the satellite to the jammer. We can calculate this angle from the law of sines for spherical triangles:

$$\sin B = (\sin b)(\sin A)/(\sin a) = \sin(45°)(\sin(2°))/\sin(5.73°)$$
$$= (.707)(.0350)/(.0998) = .247$$

Angle B = 14.4°.

Now consider Figure 7.20. This is a plane triangle formed by the satellite, the center of the Earth and the jammer location. Side d is the range from the satellite to the jammer. Side e is the radius of the Earth (R_E) plus the elevation of the satellite (H). Side f is the radius of the Earth. Angle D is the same angle we calculated as a in the spherical triangle of Figure 7.19.

Angle D is 5.73°. This is the same as side a in the spherical triangle of Figure 7.19.

Side e is 6,671 km.

Side f is 6,371 km.

The law of cosines for sides in plane triangles is:

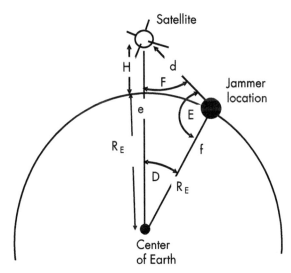

Figure 7.20 The propagation distance between a receiving satellite and a jamming transmitter on the Earth's surface can be calculated from the plane triangle formed by the satellite location, the jammer location, and the center of the Earth.

$$d^2 = e^2 + f^2 - 2ef\cos D$$
$$= (6{,}671 \text{ km})^2 + (6{,}371 \text{ km})^2 - 2(6{,}671)(6371)\cos 5.73°$$
$$= 44{,}502{,}241 + 40{,}589{,}641 - 84{,}557164 \text{ km}^2$$
$$= 534718 \text{ km}^2$$
$$d = \text{sqrt}(534718 \text{ km}^2)$$
$$= 731 \text{ km}$$

7.5.2 Jamming Link Loss

The propagation loss in the jamming link is the LOS:

$$\text{Loss}_J = 32.4 + 20\log(d) + 20\log(F) = 32.4 + 20\log(731) + 20\log(5000)$$
$$= 32.4 + 57.2 + 74 = 163.6 \text{ dB}$$

Still in Figure 7.20, we can determine angle F from the law of sines for plane triangles:

$$\sin F = f \sin D / d = 6{,}371 \sin(5.73°) / 731 = (6{,}371)(.0998)/(731) = .871$$

Angle F = 60.6°.

7.5.3 Gain of 4-m Jamming Antenna at 5 GHz

The antenna boresight gain can be determined from the formula:

$$G = -42.2 + 20\log D + 20\log F$$

where G is the boresight gain in dBi, D is the diameter of the antenna in meters, and F is the operating frequency in megahertz.

For the jammer antenna,

$$G = -42.2 + 20\log(4) + 20\log(5000)$$
$$G = -42.2 + 12 + 74 = 43.8 \text{ dBi}$$

7.5.4 Jammer ERP

The ERP of the jammer is 50 dBm + 43.8 dBi = 103.8 dBm.

7.5.5 The Satellite Uplink

Figure 7.21 is a spherical triangle formed by the satellite SVP, the ground station location, and the North Pole. Side g is the geocentric angle between the satellite SVP and the ground station. Side k is 90° minus the latitude of the SVP. Side j is 90° minus the latitude of the ground station. Angle G is the difference in longitude between the subvehicle location and the ground station.

Side k = 90° − 40° = 50°
Side j = 90° − 42° = 48°
Angle G = 3°

The spherical law of cosines for sides is:

$$\cos g = (\cos b)(\cos c) + (\sin b)(\sin c)(\cos A)$$
$$= (\cos 50°)(\cos 48°) + (\sin 50°)(\sin 48°)(\cos 2°)$$
$$= (0.643)(0.669) + (0.766)(0.743)(0.999)$$
$$= 0.430 + 0.569 = 0.999$$

So side g = 2.83°.

Angle J in Figure 7.21 is the azimuth from the satellite to the ground station. We can calculate this angle from the law of sines for spherical triangles:

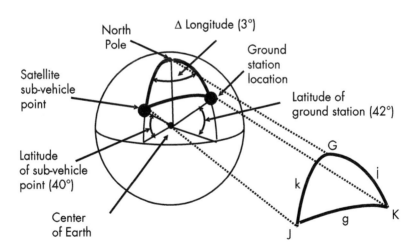

Figure 7.21 A spherical triangle is formed between the North Pole, the satellite SVP, and the ground station location.

$$\sin J = (\sin j)(\sin G)/(\sin g) = \sin(48°)(\sin(3°)/\sin(2.83°))$$
$$= (.743)(.0323)/(.0492) = .488$$

Angle $J = 29.2°$.

7.5.6 Gain of 2-m Uplink Transmitter Antenna

$$G = -42.2 + 20\log D + 20\log F$$

For the satellite uplink transmit antenna,

$$G = -42.2 + 20\log(2) + 20\log(5,000)$$
$$G = -42.2 + 6 + 74 = 37.8 \text{ dBi}$$

7.5.7 Uplink ERP

The ERP of the uplink transmitter is 40 dBm + 37.8 dBi = 77.8 dBm.

Now consider Figure 7.22. This is a plane triangle formed by the satellite, the center of the Earth, and the ground station location. Side p is the range from the satellite to the ground station. Side n is the radius of the Earth (R_E) plus the elevation of the satellite (H). Side m is the radius of the Earth. Angle P is the same angle that we calculated as side g in the spherical triangle of Figure 7.21.

Angle P is 2.83°. This is the same as side g in the spherical triangle of Figure 7.21.

Side n is 6,671 km.
Side m is 6,371 km.
The law of cosines for sides in plane triangles is:

$$p^2 = n^2 + m^2 - 2mn\cos P$$
$$= (6,671 \text{ km})^2 + (6,371 \text{ km})^2 - 2(6,671)(6371)\cos 2.828°$$
$$= 44,502,241 + 40,589,641 - 84,916,880 \text{ km}^2$$
$$= 175,002 \text{ km}^2$$
$$p = \text{sqrt}(175,002 \text{ km}^2)$$
$$= 418 \text{ km}$$

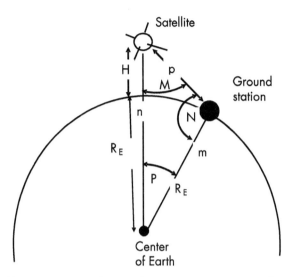

Figure 7.22 The propagation distance between a receiving satellite and its ground station on the Earth's surface can be calculated from the plane triangle formed by the satellite location, the ground station location, and the center of the Earth.

7.5.8 Uplink Loss

The loss in the satellite uplink is the LOS:

$$Loss_S = 32.4 + 20\log(418) + 20\log(5{,}000) = 32.4 + 52.4 + 74 = 158.8 \text{ dB}$$

7.5.9 Gain and Bandwidth of the Uplink Receiving Antenna on the Satellite

The boresight gain of the antenna is:

$$G = -42.2 + 20\log D + 20\log F$$
$$G = -42.2 + 20\log(1) + 20\log(5{,}000)$$
$$G = -42.2 + 0 + 74 = 31.8 \text{ dB}$$

The average sidelobe gain is 20 dB below the boresight gain.

7.5.10 Antenna 3-dB Beamwidth

$$\alpha = \text{antilog}\big[(86.8 - 20\log D - 20\log F)/20\big]$$

where α is the 3-dB beamwidth in degrees, D is the diameter of the antenna in meters, and F is the operating frequency in megahertz.

$$\alpha = \text{antilog}\big[(86.8 - 20\log(1) - 20\log(5{,}000))/20\big]$$
$$= \text{antilog}\big[(86.8 - 0 - 74)/20\big] = 4.4°$$

Still, in Figure 7.22, we can determine angle M from the law of sines for plane triangles:

$$\sin M = m\sin P / p = 6371\sin(2.82°)/418 = (6371)(.0492)/(418) = .750$$

Angle M = 48.3°.

7.5.11 Offset of the Jammer from the Downlink Receiving Antenna

Now we will determine the angle between the boresight of the downlink receiving antenna and the jammer.

Figure 7.23 is a right spherical triangle with the satellite as the origin of the sphere.

Side aa is the difference in azimuth between the ground station and the jammer.

Side bb is the difference in elevation between the ground station and the jammer.

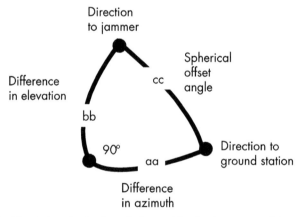

Figure 7.23 The right spherical angle formed by the jammer and the ground station can be defined by the difference in azimuth and difference in elevation, both as seen from the satellite.

7.5 JAMMING SATELLITE UPLINKS

Side cc is the satellite-central angle between the ground station and the jammer.

The uplink receiving antenna has its boresight aimed at the ground station, so the jammer is in a sidelobe offset from the boresight by angle cc.

In this problem, the azimuth from the satellite to the jammer is 14.4° from Figure 7.19. The azimuth from the satellite to the ground station is 48.3 from Figure 7.21. The difference between the azimuth values is 19.5°. This is side aa in the spherical triangle of Figure 7.23. The elevation from the satellite to the jammer is 60.6° from Figure 7.20. The elevation from the satellite to the ground station is 48.3° from Figure 7.22. The difference between the elevation values is 12.3°. This is side bb in the spherical triangle of Figure 7.23.

From Napier's rules for right spherical triangles,

$$\cos(\text{side } cc) = [\cos(\text{side } aa)][\cos(\text{side } bb)] = [\cos(19.5°)][\cos(12.3°)]$$
$$= [.923][.977] = .902$$

Angle cc = 25.5°.

Since this offset angle is much wider than the 4.4° receiving antenna beamwidth, the gain of the uplink receiving antenna toward the jammer will be reduced to its average sidelobe level, which is 31.8 dB − 20 dB = 11.8 dBi.

7.5.12 The J/S

The formula for communications jamming is

$$J/S = ERP_J - ERP_S - LOSS_J + LOSS_S + G_{SJ} - G_S$$

where J/S is the jamming-to-signal ratio in decibels, ERP_J is the ERP of the jammer in dBm (103.8 dBm), ERP_S is the ERP of the jammed signal in dBm (77.8 dBm), $LOSS_J$ is the loss between the jammer and the jammed receiver in decibels (163.6 dB), $LOSS_S$ is the loss between the desired signal transmitter and the jammed receiver in decibels (158.8 dB), G_{SJ} is the gain of the jammed receiver's antenna in the direction of the jammer in decibels (11.8 dB), and G_S is the gain of the jammed receiver's antenna in the direction of the desired signal transmitter in decibels (31.8 dB).

Plugging in the values for this problem:

$$J/S = 103.8 \text{ dBm} - 77.8 \text{ dBm} - 163.6 \text{ dB} + 158.8 \text{ dB}$$
$$+ 11.8 \text{ dB} - 31.8 \text{ dB} = 1.2 \text{ dB}$$

This is effective jamming assuming that frequency modulation (FM) noise jamming modulation is used.

7.6 ELECTRONIC PROTECTION OF SATELLITE LINKS

7.6.1 Attacks on Links

Satellite links are vulnerable to intercept, jamming, and spoofing. Therefore, it is important to protect those links from hostile activity.

7.6.1.1 Intercept

Intercepting a link means receiving it by someone other than the intended receiver. Sometimes the hostile receiver may just be interested in learning the signal externals such as transmission frequency, modulation, emitter location, encryption approach, or timing issues. This is called electronic support.

However, sometimes the intention of the hostile receiver is to recover the internals of the signals, that is, the information they carry. This is called communications intelligence (COMINT).

Intercept can be prevented by the same techniques described below for protection against jamming. Also, the use of encryption can prevent the recovery of signal internals.

7.6.1.2 Spoofing

If false commands are transmitted to a satellite and are accepted as valid, they can cause the satellite or its payload to perform functions that are hostile to the proper operation of the system. In some cases, they can disable the satellite or end its mission. This can be overcome by authentication measures or by the techniques described below to overcome jamming.

7.6 ELECTRONIC PROTECTION OF SATELLITE LINKS

7.6.1.3 Link Jamming

Since satellite links are typically digital, jamming them can involve preventing the link from synchronizing or by creating bit errors.

Links must carry their information in a serial format, so it is necessary to synchronize the receiver with the transmitter in order to recover the information. If it is practical to prevent synchronization, jamming can be very efficient. However, the synchronization of all but a few commercial links can be expected to be very robust. Therefore, it is most practical to jam the link by causing it to generate large numbers of bit errors.

Digital bits cannot be directly transmitted; they must first be modulated onto radio frequency carriers with ones and zeros differentiated by frequency, amplitude, or, most frequently, phase.

Bit errors are transmitted ones received as zeros or vice versa. There is a relationship between the predetection signal-to-noise ratio in the received signal and the percentage of bit errors that are output. This ratio depends on the modulation scheme used. The relationship between the modulation and the bit errors is illustrated in Figure 7.24. As a general rule, if the bit error rate is about 25%, communication

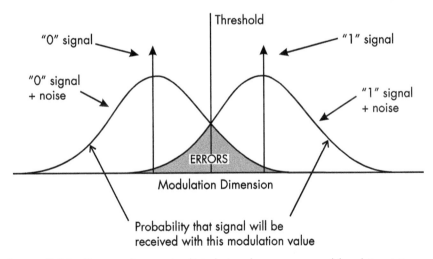

Figure 7.24 Ones and zeros in digital signals are recovered by determining on which side of a threshold they occur in the received signal. The signal + noise envelope is the probability that a one or zero will be received at that point in the modulation dimension. Bit errors occur when the noise takes the signal to the wrong side of the threshold.

cannot take place. This bit error rate must be present in every syllable of voice communication or in every subframe of digital data. A generally used criterion for the creation of enough bit errors to stop communication is that the predetection J/S be at least unity (i.e., 0 dB).

7.6.2 Protection Against Jamming

There are several ways to protect against the creation of bit errors by jamming; they include error correction codes, majority encoding, and spectrum spreading.

Error correction codes add extra bits to each block of digital data that allow the receiver to detect and correct the errors detected up to some maximum error percentage. Since these extra bits require greater signal bandwidth, they will make the receiver less sensitive and thus require stronger link signals.

Majority encoding involves repeating data blocks several times and comparing the received blocks before selecting the block with the maximum correlation. This, again, requires extra signal bandwidth.

The most common way to protect satellite links is by spectrum spreading. This involves adding a secondary modulation to a digital signal so that it is transmitted over a wider bandwidth. When the spread signal is received by an authorized receiver, the spreading code is removed so that the correct original modulation is output. Unauthorized receivers will not be able to remove the spreading code as shown in Figure 7.25. This means that the unauthorized receiver must have significantly wider bandwidth, usually orders of magnitude. Thus, the received signal-to-noise ratio is significantly reduced and intercept or jamming becomes problematic. Note that there are some techniques that can be used by very sophisticated hostile systems to read through the spreading with less loss of signal quality.

There are two common spreading approaches used in satellite links: frequency hopping and direct sequence spread spectrum. Either or both can be used to protect links.

7.6.2.1 Frequency Hopping

As shown in Figure 7.26, frequency hopping involves changing frequency after each block of many bits. This is a very mature approach and is widely used. It allows very wide frequency spreading; how-

7.6 ELECTRONIC PROTECTION OF SATELLITE LINKS

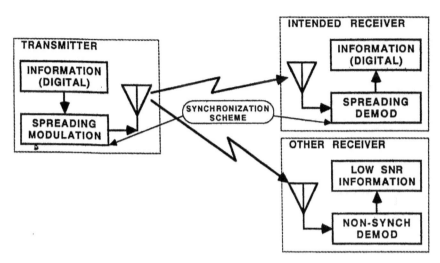

Figure 7.25 Communication electronic protection prevents unauthorized receivers from receiving protected signals. There are several types of spreading modulation. The most common for satellite links are frequency hopping or direct sequence spread spectrum.

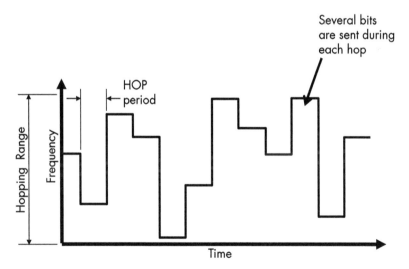

Figure 7.26 A frequency-hopped signal changes frequency after each several bits are sent.

ever, it is the easiest spreading scheme to defeat using sophisticated modern software-driven techniques.

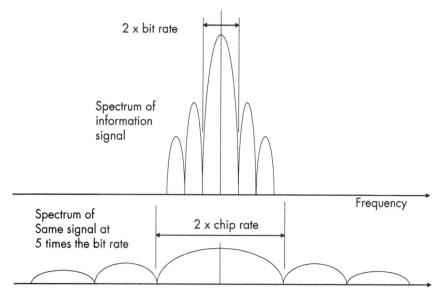

Figure 7.27 Frequency spreading versus bit rate. Chips are bits used to spread the signal, and bits carry the information.

7.6.2.2 Direct Sequence Spread Spectrum

This involves applying a second digital modulation that has a significantly higher bit rate. As shown in Figure 7.27, the transmitted power is spread over a wider spectrum by the ratio of the increase in the modulation bit rate. This figure represents an under-exaggeration in that the spreading modulation is only five times as fast as the information carrying modulation. In general, the signal is spread by two or three orders of magnitude. This is considered the most secure way to protect satellite links against EW attacks.

8

DURATION AND FREQUENCY OF OBSERVATIONS

In this chapter, we will first calculate the distance over which a satellite can "see" as a function of its orbital parameters. Then we will calculate how long the satellite can see that point and, finally, the frequency shift in signals received by the satellite or an Earth-based receiver. All three of these calculations are important to the application of EW functions involving satellites. All of these calculations are dependent on the location of the satellite and the location of the Earth surface point being considered.

In Chapter 6, Section 6.1.2, we calculated the look angles from a satellite with a 1,000-km-high circular orbit with a 3-hour (180-minute) period and an SVP at 30° North latitude and 100° East longitude. The target that we are considering is on the surface of the Earth at 45° North latitude and 120° East longitude.

From Table 8.1 (which is the same as Table 3.2, reprinted here for convenience), we know that this satellite will have a semi-major axis of 10,560 km. Since it has a circular orbit, it will be at a constant altitude of 4,189 km.

Figure 8.1 shows a satellite and the Earth surface area that it can see. In this figure, there has been no attempt to draw the satellite altitude to scale; we will handle that in the math. The higher the satellite, the more of the Earth's surface is available for the intercept or jamming of targets.

Table 8.1
Altitude and Semi-Major Axis of Circular Orbits Versus the Satellite Period

p(min)	h(km)	a(km)	p(min)	h(km)	a(km)
90	281	6652	330	9447	15818
105	1001	7372	345	9923	16294
120	1688	8059	360	10392	16763
135	2346	8717	375	10854	17225
150	2980	9351	390	11311	17682
165	3594	9965	405	11761	18132
180	4189	10560	420	12206	18577
195	4768	11139	435	12646	19017
210	5332	11703	450	13081	19452
225	5883	12254	465	13510	19881
240	6422	12793	480	13936	20307
255	6949	13320	495	14357	20728
270	7466	13837	510	14773	21144
285	7974	14345	525	15186	21557
300	8473	14844	540	15595	21966
315	8964	15335	—	—	—

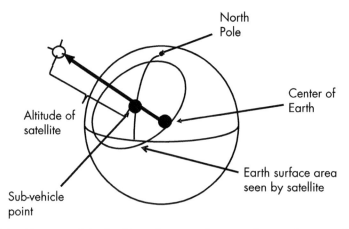

Figure 8.1 The area of the Earth's surface seen by a satellite is a function of the SVP and the altitude of the satellite.

8.1 CALCULATING THE DISTANCE TO THE HORIZON

Consider Figure 8.2. This shows a plane triangle in the plane defined by the satellite, the center of the Earth, and the most distant point on the Earth's surface that the satellite can see. The local horizon plane

8.1 CALCULATING THE DISTANCE TO THE HORIZON

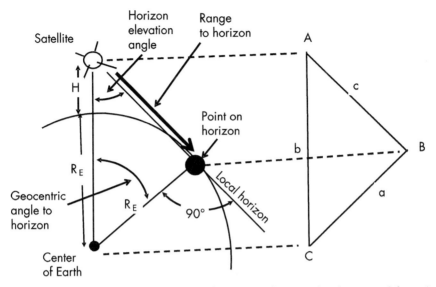

Figure 8.2 The range to the horizon from a satellite can be determined from the plane triangle defined by the satellite, a point on the horizon, and the center of the Earth.

at the chosen horizon point is tangent to the sphere of the Earth. The vector to this point from the center of the Earth intersects this plane at 90°.

As in earlier chapters, we will pull this triangle out of the diagram and label it in our normal way: that is, uppercase letters for the angles and corresponding lowercase letters for the sides opposite those angles.

Angle A is the elevation angle from the SVP to the horizon. Angle B is 90°. Angle C is the geocentric angle from the satellite to the point on the horizon. Side a is the radius of the Earth. Side b is the distance from the center of the Earth to the satellite (the height of the satellite + the radius of the Earth). Side c is the distance from the satellite to the point on the horizon.

Side a is the radius of the Earth (6,371 km), and side b is the height of the satellite above the center of the Earth (10,560 km).

Since this is a plane right triangle, the square of side b is the sum of the squares of sides a and c. So

$$c = \text{sqrt}\left[b^2 - a^2\right]$$
$$= \text{sqrt}(111{,}513{,}600 - 40{,}589{,}641) = \text{sqrt}(70{,}923{,}959) = 8{,}422 \text{ km}$$

This is the propagation distance between the satellite and a transmitter or receiver on the horizon.

Now we can use the law of sines for plane triangles to find angle C.

$$\sin C / c = \sin(B) / b$$
$$C = \arcsin\left[(c \sin B)/b\right]$$
$$= \arcsin\left[(8{,}422(\sin 90°))/10{,}560\right]$$
$$= \arcsin[8{,}422/10{,}560] = \arcsin[.7975] = 52.89°$$

From angle C, we can determine the Earth surface distance from the SVP to the horizon. The great circle circumference of the Earth is 40,030 km. Thus, we can determine the Earth surface range to the horizon along a great circle path from the formula:

$$\text{Distance} = 40{,}030 \text{ km}(\text{Geocentric Angle}/360°)$$

The elevation angle (from the nadir) to the horizon is 90° less the geocentric angle C that we just calculated.

Now we can calculate the Earth surface distance to the horizon from:

$$40{,}030 \text{ km}(\text{Geocentric Angle}/360°) = 40{,}030(52.89°/360°) = 5{,}881 \text{ km}$$

8.2 HORIZON DISTANCES FOR CIRCULAR ORBITS

Table 8.2 shows the distance to the horizon for circular satellites with various values of orbit period (in minutes). The first column is the orbital period in minutes, the second column is the altitude of the satellite (h) if it has a circular orbit, the third column shows the semi-major axis (a) (for any orbit shape), the fourth column shows the direct line range (rng) to the horizon in kilometers, and the fifth column

Table 8.2
Height, Semi-Major Axis, and Range to Horizon and Earth Surface Distance to the Horizon for Circular Satellites with the Orbital Period Specified

p(min)	h(km)	a(km)	rng(km)	dist(km)	p(min)	h(km)	a(km)	rng(km)	dist(km)
90	281	6,652	1,914	1,859	330	9,447	15,818	14,478	7,365
105	1,001	7,372	3,710	3,359	345	9,923	16,294	14,997	7,447
120	1,688	8,059	4,935	4,198	360	10,392	16,763	15,505	7,523
135	2,346	8,717	5,950	4,785	375	10,854	17,225	16,004	7,593
150	2,980	9,351	6,845	5,232	390	11,311	17,682	16,494	7,658
165	3,594	9,965	7,662	5,587	405	11,761	18,132	16,976	7,719
180	4,189	10,560	8,422	5,880	420	12,206	18,577	17,451	7,776
195	4,768	11,139	9,137	6,127	435	12,646	19,017	17,918	7,830
210	5,332	11,703	9,817	6,339	450	13,081	19,452	18,379	7,880
225	5,883	12,254	10,467	6,523	465	13,510	19,881	18,833	7,928
240	6,422	12,793	11,093	6,685	480	13,936	20,307	19,281	7,973
255	6,949	13,320	11,698	6,829	495	14,357	20,728	19,724	8,016
270	7,466	13,837	12,284	6,958	510	14,773	21,144	20,162	8,056
285	7,974	14,345	12,853	7,075	525	15,186	21,557	20,594	8,095
300	8,473	14,844	13,408	7,180	540	15,595	21,966	21,021	8,131
315	8,964	15,335	13,949	7,277	—	—	—	—	—

shows the Earth surface distance (*dist*) from the SVP to the horizon in kilometers.

8.3 CALCULATING THE DURATION OF TARGET AVAILABILITY FROM A SATELLITE

In Section 8.1, we talked about the portion of the Earth that can be seen by a satellite. Now we are going to talk about some timing issues. This gets a little complicated because the satellite is moving around the Earth while the Earth is turning on its axis inside the orbit. Table 8.2 shows the elements of orbits for various orbital periods. Now we are going to deal with two more of these elements: the inclination of the orbit (the angle at which it crosses the Equator) and the ascending node (the longitude at which the orbit crosses the Equator).

In Figure 8.3, you can see the satellite Earth trace with the ascending node and the inclination. The satellite is shown right above

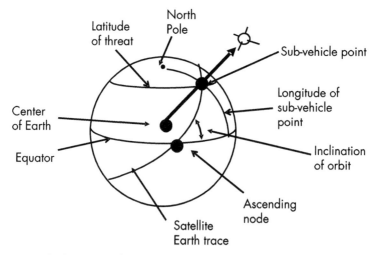

Figure 8.3 The longitude of the ascending node, the inclination, and the geocentric angle from the ascending node determine the location of the satellite SVP.

a threat location. This figure shows the Earth trace of the satellite directly over a target.

For the moment, we will stop the Earth's rotation to talk about the motion of the satellite over a nonrotating Earth. We will rotate the Earth later. Now consider Figure 8.4. When the satellite is directly over the target, the point along its path that it first rises above the horizon is F and the point at which it drops below the horizon is G.

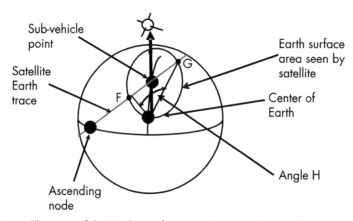

Figure 8.4 The area of the Earth's surface seen by a satellite is a function of the SVP and the altitude of the satellite.

8.3 CALCULATING THE DURATION OF TARGET AVAILABILITY

Another way to look at this is that the satellite can first see the threat when its SVP is at F and the last time that it sees the threat is when its SVP is at G.

8.3.1 Geocentric Viewing Angle

Now consider Figure 8.5. This is a spherical triangle defined by the SVP, a location on the Earth, and the North Pole. The geocentric angle between the satellite and any Earth surface location is given by the spherical triangle law of cosines for sides by the formula:

$$\cos a = (\cos b)(\cos c) + (\sin b)(\sin c)(\cos A)$$

where a is the geocentric angle between the satellite and the point on the Earth, b is 90° minus the latitude of the satellite, c is 90° minus the latitude of the point on the Earth, and A is the difference in longitude between the satellite and the point on the Earth.

Figure 8.6 is a plane triangle in the plane containing the satellite, two horizon points, and the center of the Earth. Angle G is the geocentric angle between the SVP and either one of the horizon points. It is as calculated in Section 8.1. The geocentric angle to the horizon is the arccos of the radius of the Earth/the radius of the satellite orbit.

Figure 8.7 is a view of the satellite's orbital plane. The satellite in its circular orbit has constant velocity, so the time at which it passes

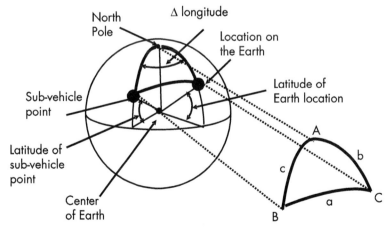

Figure 8.5 A spherical triangle is formed between the North Pole, the SVP, and location on the Earth.

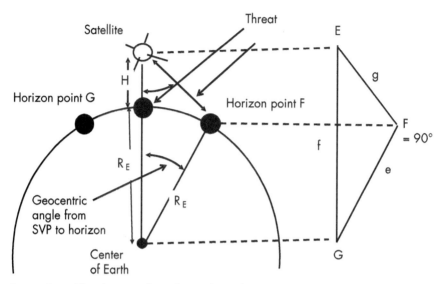

Figure 8.6 The elevation from the nadir and range to a point on the Earth from a satellite can be determined from the plane triangle defined by the satellite, the Earth location, and the center of the Earth.

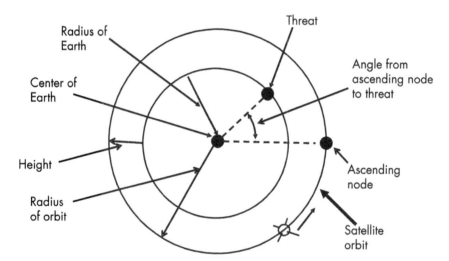

Figure 8.7 This shows the great circle plane of the satellite orbit. The angle from the ascending node increases at a constant rate.

over a threat after passing the ascending node is the period of the orbit reduced by the ratio of the geocentric angle between the ascending node and the threat divided by 360°.

8.3.2 Time During Which a Satellite Can See a Point on the Earth

Now consider Figure 8.8. This is the satellite's orbital plane, but it shows the threat along with points F and G. The geocentric angle between these two points is angle H. The time over which the satellite can see the threat (again if the Earth is not rotating) is given by the period of the satellite reduced by the angle between the two horizon points (angle H) divided by 360°. The satellite moves East during this time by a distance reduced by the cosine of the orbit inclination.

The Earth surface distance to the horizon from a satellite is the distance from the threat to either point F or point G. Thus, the satellite can see the threat during the time that it takes the SVP to travel from F to G. It is twice the arccos(radius of the Earth/the semi-major axis of the orbit). Now let's put in some numbers. The time that it takes the SVP to travel from F to G can be calculated from:

$$\text{Period of Satellite} \left(\text{Angle H} / 360° \right)$$

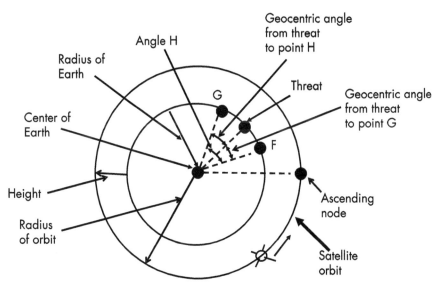

Figure 8.8 This shows the geocentric angle from the threat to each of the horizon points.

For example, if the satellite period is 180 minutes, the semi-major axis is 10,560 km. The radius of the Earth is 6,371 km, so angle H is

$$2 \times \arccos(6,371 / 10,560) = 105.78$$

Therefore, the observation time over a nonrotating Earth is:

$$180 \text{ minutes} \times (105.78 / 360) = 52.8 \text{ minutes}$$

8.3.3 The Impact of the Movement of the Earth

In the interest of avoiding spending a whole chapter on some rather abstract spherical trigonometry, we will make a pretty close estimate of the amount of additional look time that the satellite gets because of the Earth's rotation.

The time that it takes the satellite to move from point F to point G (in Figure 8.8) is increased by the amount that the observed point on the Earth moves East during that time. The observation time over the rotating Earth is longer than the observation time over a nonrotating Earth because the satellite goes over a fixed latitude and longitude point at a slower net speed. The satellite's speed over the Earth is the actual satellite speed reduced by the speed at which the Earth is moving under the satellite. If the Earth were not turning, the time that the satellite SVP would take to cover the distance from point F to point G is the satellite's period reduced by the ratio of the geocentric angle H over 360° (H/360°). During this time, the target moves East by a distance equal to the circumference of the Earth reduced by $\cos(lat)$ and further reduced by the ratio of angle H/360°.

First the math: The satellite's SVP moves along its orbit at the velocity of $2\pi\, a/P$. The eastward velocity of the satellite as it passes the target is

$$(2\pi\, a / P) \times \cos(i)\cos(lat)$$

where a is the semi-major axis of the orbit, P is the period of the satellite, i is the inclination of the orbit, and lat is the latitude of the target.

However, the Earth rotates to the East 366 times per year. This means that a spot on the Earth (e.g., the threat location) moves East at

8.3 CALCULATING THE DURATION OF TARGET AVAILABILITY

0.25068 equatorial degrees per minute. This means that the eastward speed of a point on the equator is 27.90 km/minute. As shown in Figure 8.9, the distance from the Earth's axis to the surface is reduced by the cosine of its latitude. The speed at which a point on the Earth moves East is therefore:

$$(27.90 \text{ km/min})\cos(lat)$$

8.3.4 Viewing Time Formula

The time to move East from the longitude of point F to the longitude of G is the distance divided by the velocity. The eastward distance that a point on the Earth moves is a function of latitude. The eastward movement distance for points F and G while the satellite SVP is moving between those two points is:

$$2\pi R_E \cos(lat)\cos(i)(H/360°) \text{ km}$$

The velocity at which a point on the Earth moves by this rotation is 27.90 cos(lat) km/min.

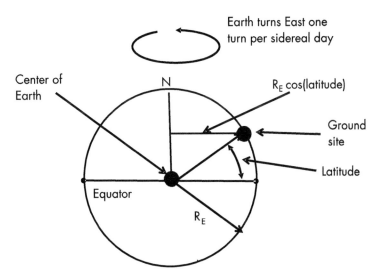

Figure 8.9 The latitude of the Earth station determines its distance from the Earth's rotation axis.

It takes the satellite 52.8 minutes to travel from point F to point G. If the inclination of the orbit is 60° and the target is at 45° latitude, this constitutes a rate toward the East of 105.78 cos(*lat*) cos(*i*)/52.8 minutes. That is a rate of 0.708° of longitude per minute.

During the time for movement from F to G (52.8 minutes), the Earth is moving East at rate of 0.25°/min, so the satellite must travel farther by 0.25°/min for 52.8 minutes for a total of 13.2° since the satellite moves East at 0.708°/min, the time that the satellite can see a threat at this spot on the Earth is extended by 18.6 minutes. Thus, the total intercept time (horizon to horizon) is 71.4 minutes.

So what is the formula for the total viewing time?

$$T_{TOTAL} = P\left[2\arccos(R_E/a)/360°\right]\left[1 + 0.25\cos(i)\cos(lat)\right]$$

8.4 DOPPLER SHIFT IN SATELLITE LINK

Since a satellite is traveling through space at a high rate of speed, you would expect that a fixed site on the Earth's surface would receive signals from the satellite at frequencies offset from the transmission frequencies by the Doppler shift. In this section, we will calculate the amount of Doppler shift as a function of the orbital parameters and the location of the receiving site on the Earth.

We will consider the maximum frequency shift of signals from a circular satellite that passes directly over a fixed Earth station. The maximum frequency shift will occur as the satellite appears at the ground station horizon.

8.4.1 Doppler Shift Formula

The Doppler shift formula gives the change in frequency as a function of the transmission frequency and the rate of change of range between the transmitter and the receiver. The satellite is following an orbital path around the Earth, and the Earth is turning inside that orbit, so both the transmitter and the receiver are moving. The Doppler shift will be a function of the difference between the two velocities adjusted for the angle between the two velocity vectors.

$$\Delta F = FV(\cos\theta)/c = FV_R/c$$

8.4 DOPPLER SHIFT IN SATELLITE LINK

where ΔF is the Doppler frequency shift, F is the transmission frequency, V is velocity of moving transmitter or receiver, c is the speed of light, θ is the true spherical angle between the velocity vector of the moving transmitter or receiver and the signal vector, and V_R is the radial velocity (i.e., the rate of change of the range between the transmitter and the receiver. Observing from the receiver, the relative velocity of the transmitter is reduced by the cosine of the angle θ.

To picture this, consider the tone of a train whistle as heard from a moving car as a moving train passes it.

8.4.2 Receiving Site Velocity

As shown in Figure 8.9 the radius of a path around the Earth is reduced by the cosine of the latitude. Thus, the ground station is moving to the East at a rate of speed equal to the Earth's rotation rate times the radius of the Earth, reduced by the cosine of the latitude angle.

If the ground station is at 30° North latitude, the distance from the Earth's axis to the ground station is the radius of the Earth reduced by the cosine of 30°. The distance that the ground site travels in a sidereal day is 2π times this reduced radius. A sidereal day (the time for one rotation of the Earth in space) is 86,164 seconds.

Cosine of 30° = 0.866

Thus, the speed of the ground station (V_E) is:

$$V_E = 2\pi (6,371 \text{ km})(0.866)/86,164 \text{ km/sec}$$
$$= 402.3 \text{ m/sec (to the East)}$$

8.4.3 Satellite Velocity

As shown in Figure 8.10, the satellite is moving along its orbital path at a speed calculated from the semi-major axis of the orbit and the satellite's period. Let's choose a circular satellite with a period of 3 hours. From Table 8.1, we see that this makes the height above the Earth equal to 4,189 km and the semi-major axis equal to 10,560 km. This means that the satellite travels through one orbit (2π)10,560 km in 3 hours (10,800 seconds). This makes the speed of the satellite (V_S) equal to:

$$V_S = 2\pi (10,560) \text{ km}/10,800 \text{ sec} = 6144 \text{ m/sec}$$

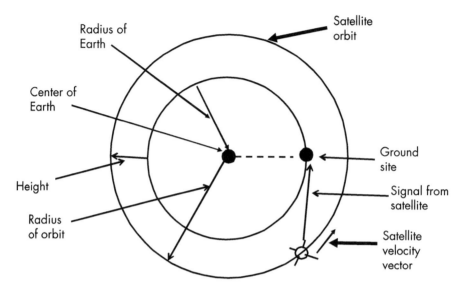

Figure 8.10 The satellite has the maximum velocity toward a ground station when it is at the horizon. Its velocity is the circumference of its orbit divided by its orbital period.

The maximum frequency shift will occur for the case in which the satellite passes directly over the ground site. Figure 8.11 shows this geometry. In the expanded spherical triangle, side a is 90° minus the Earth site latitude. Angle A is 90° minus the inclination of the satellite orbit. Angle B is 90° minus the angle North of East at which the satellite passes over the Earth station. Side b is 90°. From the law of sines for spherical triangles:

$$\sin B = (\sin A \times \sin b) / \sin A$$

So the sine of angle B in this figure is the sine of 90° minus the orbital inclination divided by the sine of 90° minus the latitude of the ground station. (Side b is 90°, so $\sin b = 1$.)

With the ground station at 30° latitude and the inclination equal to 45°, this means that angle B in the figure is 54.7°, so the satellite will pass over the Earth station in the direction 35.3° North of East (i.e., 90 degrees–angle B). Note that the inclination must be at least equal to the latitude of the ground site for the satellite to pass directly overhead.

8.4 DOPPLER SHIFT IN SATELLITE LINK

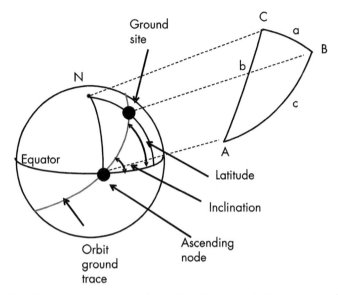

Figure 8.11 The spherical triangle formed by the ground site, the ascending node, and the North Pole allows calculation of the direction that the satellite ground trace passes through the location of the Earth station.

Figure 8.12 shows the horizon plane at the ground site. In this figure, the angle between the velocity vector of the ground site and the velocity vector of the satellite is 90° minus angle B from Figure 8.11 (i.e., 35.3°).

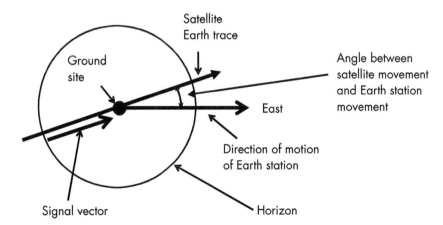

Figure 8.12 This view of the horizon plane of the Earth station shows the velocity vector of the satellite and of the Earth station motion with the Earth's rotation.

The satellite signal will have the maximum positive frequency shift as it appears at the horizon and the maximum negative frequency shift as it disappears over the horizon. If we take the horizon area to be a small fraction of the Earth's surface, the direction of arrival of the signal from the satellite is aligned with the velocity vector (i.e., 35.3° from the East).

The rate of change of range is now $V_S - (V_E \cos 35.3°)$. For this case, the rate of change of range is:

$$6{,}144 \text{ m/sec} - 402.3(\cos 35.3°) \text{m/sec} = (6{,}144 - 328) \text{ m/sec} = 5{,}816 \text{ m/sec}$$

If the signal from the satellite is at 5 GHz, the Doppler shift from the formula at the beginning of this section is:

$$\Delta F = F(V_R / c)$$
$$= 5{,}000 \ MHz (5816 / 3 \times 10^8) = 96.9 \ kHz$$

8.4.4 General Formula for Maximum Doppler Shift

After all of the above discussion, here is an ugly formula for the maximum Doppler shift for a signal to or from a satellite passing directly over a ground station:

$$\Delta F = [F / 1.8 \times 10^7][(2\pi a / P) - 27.9 \cos(i)]$$

where ΔF is the frequency shift in megahertz, F is the transmission frequency in megahertz, a is the radius of the circular orbit in kilometers, P is the period of the orbit in minutes, and i is the inclination of the orbit in degrees.

Note that the speed of light is converted from the familiar 3×10^8 m/s to 1.8×10^7 km/min.

8.4.5 General Formula for the Doppler Shift

Whether or not the satellite passes over the ground receiver, the following discussion allows the calculation of the Doppler shift in either

8.4 DOPPLER SHIFT IN SATELLITE LINK

the uplink or downlink between a satellite and a ground-based receiver at any point in the satellite's orbit.

This discussion starts with the azimuth and elevation to a satellite from a ground receiver location as shown in Figure 8.13 relative to the velocity vector of the receiver.

The purpose is to determine the velocity of the receiver toward the transmitter. In the spherical triangle of the figure, side a is the azimuth to the transmitter and side b is the elevation. Using Napier's rules:

$$\cos(c) = \cos(a)\cos(b)$$

Since both of these angles are relative to the receiver's velocity vector, cos(c) is Θ_1 in the formula below. This is the factor by which the velocity of the receiver must be reduced when calculating the Doppler shift.

Now consider Figure 8.14. This spherical triangle has the azimuth from the transmitter to the receiver as side d and the elevation as side e. Again by Napier's rules, the cosine of side f is the product of the cosines of sides d and e. It is Θ_2 in the formula below.

$$\cos(f) = \cos(d)\cos(e)$$

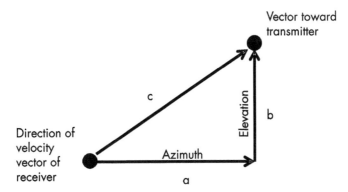

Figure 8.13 This right spherical triangle shows the location of the transmitter relative to the receiver velocity vector.

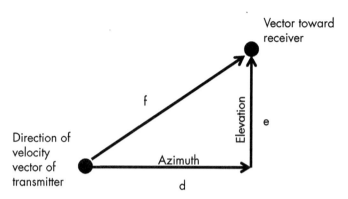

Figure 8.14 This right spherical triangle shows the location of the receiver relative to the transmitter velocity vector.

This is the factor by which the velocity of the transmitter must be reduced when calculating the Doppler shift.

The formula for the Doppler shift for either an uplink or a downlink is given by the formula:

$$\Delta F = F\left[(V_R \cos \Theta_1 + V_T \cos \Theta_2)/c\right]$$

where ΔF is the Doppler shift, F is the transmitted frequency, V_R is the velocity of the receiver, Θ_1 is the true spherical angle from the receiver's velocity vector to the transmitter, V_T is the velocity of the transmitter, and Θ_2 is the true spherical angle from the transmitter's velocity vector to the receiver. c in this equation is the speed of light.

Relating this formula to the earlier discussion, the velocity vector of the transmitter or receiver that is on the Earth is 27.9 km/min directly East. The speed of the transmitter or receiver in the satellite is the circumference of the orbit divided by the orbital period. The direction of the satellite's velocity vector changes constantly as the satellite passes by the ground location. The part of the formula $V_R \cos \Theta_1 + V_T \cos \Theta_2$ equals the rate of change of range between the transmitter and the receiver.

9

INTERCEPT FROM SPACE

In the earlier chapters, we discussed orbit mechanics, satellite links, radio propagation, and link vulnerability. Now we will dig into the EW applications of these subjects. The two primary EW tasks that a satellite is asked to perform are intercept and jamming. In this chapter, we will cover the intercept of hostile signals by receiving systems in satellites. We will consider intercept missions from both low Earth satellites and synchronous satellites.

Please be aware that the problems that we consider in this chapter use arbitrarily chosen satellite configurations, orbit parameters, and target capabilities. All of these values are supposed to be reasonable. They do not represent any existing real-world systems. The idea is for you to see how the equations apply and the results that they produce. Later, when you need to analyze a real-world problem, you can plug in the real-world values to get the real-world answers.

9.1 INTERCEPT OF RADAR SIGNAL FROM LOW-EARTH SATELLITE

Here is a very practical example of an intercept. We are intercepting a radar on the Earth with a satellite payload that has a wide beamwidth antenna that can see the whole Earth surface area that is above the horizon from the satellite.

The satellite is in a circular orbit, 300 km above the Earth. The radius of the Earth is 6,371 km, so the semi-major axis of the satellite orbit is 6,671 km. The inclination of the orbit is 60°. Since this is a circular orbit, the semi-major axis is actually the radius of the orbit. The satellite's SVP is at 100° East longitude and 45° North latitude. The satellite payload is a receiving system with bandwidth = 10 MHz and Noise figure = 3 dB. It has a wide-beam, circularly polarized antenna that has 3-dB gain.

The threat signal is a 6-GHz radar with ERP = 120 dBm, antenna boresight gain = 30 dBi and average sidelobe level is 20 dB below the main beam boresight gain. This radar is on the Earth at 95° East longitude and 35° North latitude.

9.1.1 Link Loss

To calculate the link loss, we will need to know the satellite to radar path length and the elevation angle to the satellite from the threat emitter location. Figure 9.1 is a spherical triangle formed by the North Pole, the satellite's SVP, and the location of the radar that we are jamming. Side c is 90° minus the latitude of the satellite SVP. Side b is 90° minus the latitude of the target radar location. Side a is the geocentric angle between the satellite and the target radar. Using the spherical law of cosines for sides:

$$\cos(a) = \cos(b)\cos(c) + \sin(b)\sin(c)\cos A$$

The right side of this equation is:

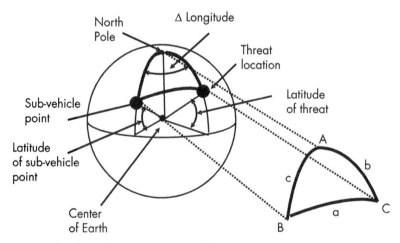

Figure 9.1 A spherical triangle is formed between the North Pole, the SVP, and the threat location.

9.1 INTERCEPT OF RADAR SIGNAL FROM LOW-EARTH SATELLITE

$$= \sin(\text{lat of SVP})\sin(\text{lat of Threat})$$
$$+ \cos(\text{lat of SVP})\cos(\text{lat of Threat})\cos(\Delta \text{ Long})$$

Plugging in our given values:

$$\cos(a) = \sin(45°)\sin(35°) + \cos(45°)\cos(35°)\cos(5°)$$
$$= (.707)(.574) + (707)(.819)(.996) = .983$$

So the geocentric angle between the satellite and the target radar is arccos(0.983) = 10.58°.

Figure 9.2 is a plane triangle formed by the satellite, the target radar, and the center of the Earth. The sides of this triangle are: f = the semi-major axis of the satellite, e = the radius of the Earth, and g = the distance from the satellite to the target radar. Angle G is the geocentric angle that we just calculated as a in the spherical triangle above. Angle F is the angle from the center of the Earth to the satellite as seen from the target radar location. Note that angle F is 90° + the elevation angle of the satellite as seen by the radar.

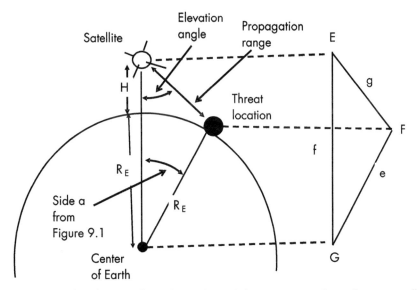

Figure 9.2 The elevation from the nadir and the range to a threat from a satellite can be determined from the plane triangle defined by the satellite, the threat, and the center of the Earth.

We can calculate the satellite to target radar range (side g) from the (plane triangle) law of cosines:

$$g^2 = e^2 + f^2 - 2ef\cos G$$
$$= (6{,}371)^2 + (6{,}671)^2 - 2(6{,}371)(6{,}671)\cos(10.58°) = 1{,}535{,}074$$

So the range from the satellite to the radar is 1,239 km.

Now we can find the satellite elevation angle as seen from the radar (angle F − 90°).

Use the (plane triangle) law of sines:

$$\sin F = f\sin(G)/g = 6{,}671\text{ km}\sin(10.58°)/1{,}239\text{ km}$$
$$= (6671)(.184)/1{,}239 = .991$$

Arcsin (0.991) equals either 82.31° or 97.82°.

The angle is greater than 90°, so it is 97.82°. The elevation angle of the satellite above the horizon is reduced from this by 90°, so it is 7.8°.

Now, at last, we will calculate the link losses.

9.1.2 LOS Loss

The LOS loss at 6 GHz and 1,239 km is:

$$32.44 + 20\log(F) + 20\log(d) = 32.44 + 75.56 + 61.86 = 169.86\text{ dB}$$

9.1.3 Atmospheric and Rain Loss

Now consider the atmospheric loss from Figure 9.3. From this graph, you can see that the atmospheric loss through the whole atmosphere at an elevation angle of 7.8° is 0.2 dB. Let's consider that we want to intercept this signal through heavy rain. In Figure 9.4, we see that the signal passes through the rain from the 0° isotherm to the ground. From Figure 9.5, we see that the 0° isotherm is below 3 km (with a 1% probability) at 35° latitude. From Figure 9.6, we can calculate that the distance through the rain from the 0° isotherm to the Earth is:

9.1 INTERCEPT OF RADAR SIGNAL FROM LOW-EARTH SATELLITE

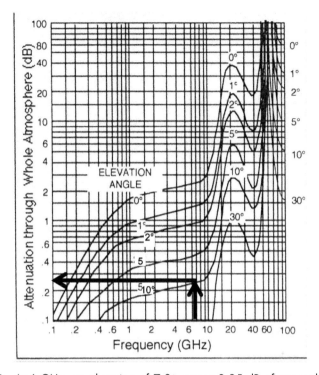

Figure 9.3 At 6 GHz, an elevation of 7.8° causes 0.25 dB of atmospheric attenuation.

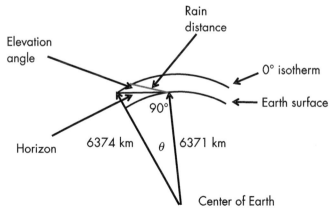

Figure 9.4 The distance through the rain is from the ground to the point at which the signal path to the satellite passes the 0° isotherm.

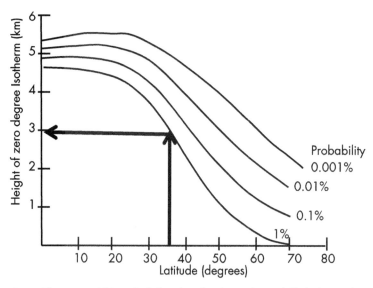

Figure 9.5 There is a 1% probability that the 0° isotherm falls below 3 km at 35° latitude.

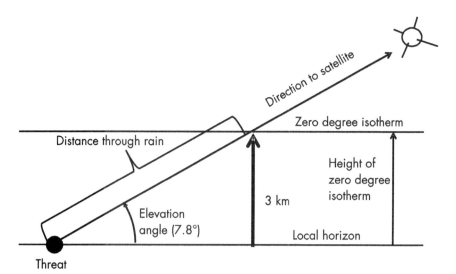

Figure 9.6 The distance over which rain loss applies is the part of the link below the 0° isotherm.

$$3 \text{ km} / \sin(7.8°) = 22 \text{ km}$$

From Figure 9.7, we see that heavy rain causes 0.1-dB loss per kilometer at 6 GHz, so the loss from rain is:

$$22 \text{ km} \times 0.1 \text{ dB} = 2.2 \text{ dB}$$

So the total link loss is 169.9 dB + 0.3 dB + 2.2 dB = 172.4 dB.

9.1.4 Can the Satellite Payload Receive the Signal?

Now we will make some additional relevant calculations. First, let's calculate the received signal strength at the satellite. The ERP of the

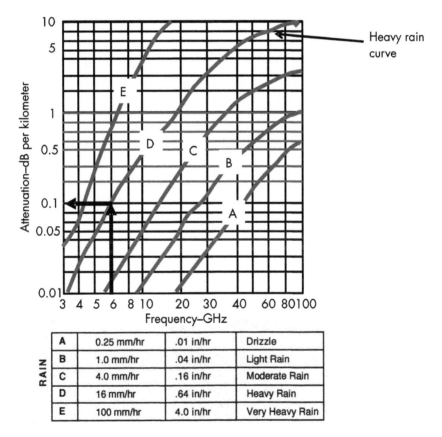

Figure 9.7 The loss from heavy rain at 6 GHz is 0.1 dB/km.

target radar is 120 dBm including an antenna boresight gain of 30 dB with an average sidelobe level 20 dB below the boresight gain. It is very unlikely that the satellite will receive the main beam from this radar. The satellite payload will see a sidelobe, so the ERP in the direction of the satellite can be assumed to be 100 dBm (20 dB below the boresight ERP). Figure 9.8 shows the signal strength through the link as a function of the above parameters. The received signal power is given by the formula:

$$P_R = ERP - L + G_R$$

where P_R is the received signal power into the satellite's receiver system, ERP is the target radar's emitted power in the direction of the satellite, L is the total link loss, and G_R is the satellite's receiving antenna gain in the direction of the target signal.

There is one additional factor in the link loss. We said the satellite antenna was circularly polarized and had a 3-dBi gain. Let's say that the antenna provides that gain over the whole Earth surface that is above the satellite's horizon. Let's assume that the target radar has a linearly polarized antenna. This means that there will be a 3-dB polarization loss in the propagation link because any circular polarization will receive any linear polarization with a 3-dB loss. There will be

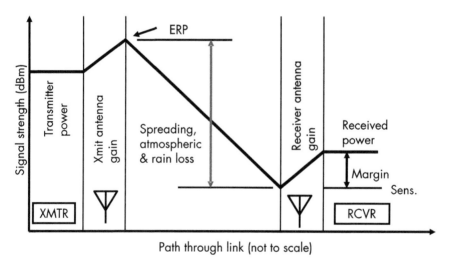

Figure 9.8 The signal strength at each point in the propagation link.

a rotation of the polarization of the radar signal as the signal passes through the atmosphere, but that will not change our link. Now, let's plug into the received power formula.

$$P_R = 100 \text{ dBm} - 172.4 \text{ }dB - 3 \text{ dB} + 3 \text{ dB} = -72.4 \text{ dBm}$$

9.1.5 Receiver Sensitivity

In setting up the problem, we stated that the satellite payload receiver had an effective bandwidth of 10 MHz and a noise figure of 3 dB. Let's assume that there is a processor on board that will analyze any received signals. This means that the receiver must provide a predetection signal-to-noise ratio of about 15 dB for the processor to provide accurate analysis results.

The receiver sensitivity is given by:

$$S = kTB + NF + RFSNR$$

where S is the receiver system sensitivity in dBm, kTB is the thermal noise in the receiver in dBm, NF is the receiver system noise figure in decibels, and $RFSNR$ is the predetection signal-to-noise ratio in decibels.

kTB is -114 dBm $+ 10\log(\text{bandwidth in MHz}) = -104$ dBm
$S = -104$ dBm $+ 3$ dB $+ 15$ dB $= -86$ dBm

9.1.6 Link Margin

The link margin, as shown in Figure 9.8, is:

$$P_R - S = -72.4 \text{ dBm} - (-86 \text{ dBm}) = 13.6 \text{ dB}$$

So the satellite payload will do a very good job of intercepting the target radar signal.

9.1.7 Could the Satellite Receive the Signal from Its Horizon?

Consider the plane triangle in Figure 9.9 formed by satellite, the target emitter at the horizon point, and the center of the Earth. Side b is

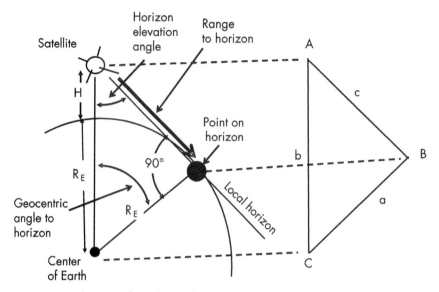

Figure 9.9 The range from the satellite to the horizon and the geocentric angle to the horizon can be calculated from the plane triangle formed by the satellite, the horizon point, and the center of the Earth.

the semi-major axis of the satellite, which is the radius of the Earth + the satellite elevation ($RE + H$). Side a is the radius of the Earth (RE). Angle C is the geocentric angle from the satellite to the horizon. It can be calculated from the formula:

$$C = \arccos[R_E / R_E + H)] = \arccos[6371/6671] = \arccos(.955) = 17.2°$$

The Earth surface distance from the satellite SVP to the horizon is:

$$(C/360°) \times 2\pi \times 6371 \text{ km} = 1913 \text{ km}$$

The link distance is found from the equation:

$$c = (R_E + H)\sin(C) = 6671\sin(17.2°) = 1973 \text{ km}$$

The spreading loss is now:

$$32.44 + 20\log(F) + 20\log(d) = 32.44 + 75.56 + 65.90 = 173.9 \text{ dB}$$

9.1 INTERCEPT OF RADAR SIGNAL FROM LOW-EARTH SATELLITE

The atmospheric loss from Figure 9.10, with the input values of 6 GHz and 0° elevation angle, is 2.2 dB.

We know that the 0° isotherm is expected to be at 3-km altitude. The distance through the rain is the distance from the ground to the point at which the signal path is 3 km above the Earth as shown in Figure 9.11. The geocentric angle (θ) is:

$$\arccos(6371/6374) = 1.76°$$

The link distance through the rain is then:

$$6374 \text{ km} \times \sin(1.76°) = 196 \text{ km}$$

This much distance through heavy rain would cause almost 20 dB of attenuations (and a serious flood), so we have to question whether this is likely, or if it would be more likely that the heavy rain

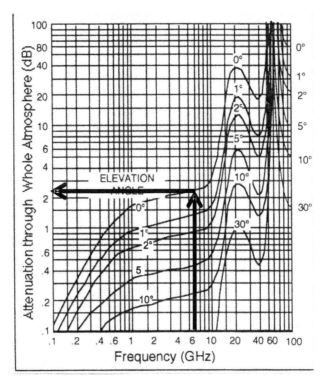

Figure 9.10 At 6 GHz, an elevation of 0° causes 2 dB of atmospheric attenuation.

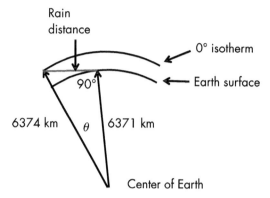

Figure 9.11 The distance through the rain is from the ground to the point at which the signal path to the satellite passes the 0° isotherm.

would be concentrated in rain cells covering 20% of the distance. If the rain cells cover 20% of the above distance, the rain attenuation would be about 3.9 dB.

Thus, the total link loss is:

$$173.9 \text{ dB} + 2.2 \text{ dB} + 3.9 \text{ dB} = 180 \text{ dB}$$

So the received power in the satellite receiver is:

$$100 \text{ dBm} - 180 \text{ dB} = -80 \text{ dBm}$$

making the operating margin equal to:

$$-80 \text{ dBm} - (-86 \text{ dBm}) = 6 \text{ dB}$$

This means that the satellite receiver will still see the target signal with 6 dB margin.

9.1.8 How Long Will the Satellite See the Signal?

With an elevation of 300 km (an orbit radius of 6,671 km), the period of the satellite will be determined from the formula:

$$P^2 = a^3 / C$$

The constant C is 3.64×10^7, so the period of this 300-km-high circular satellite is 90.3 minutes.

There is an admittedly ugly formula in Section 8.3 that gives the viewing duration as a function of the period, the inclination of the orbit, and the latitude of the emitter on the Earth's surface. Note that the observation time described in section 8.3 is making the point that the Earth moves, thus extending the viewing time. thus, it is called T sub Total. This total time is the observation time discussed here.

$$T_{TOTAL} = P\left[2\arccos(R_E / a)\right]\left[1 + \cos(i)\cos(lat)\right]$$

where T_{OB} is the time that a specific point on the Earth can be observed from a satellite, a is the satellite semi-major axis (i.e., the radius of a circular orbit), i is the satellite inclination, lat is the latitude of the ground target, R_E is the radius of the Earth, and P is the satellite's period in minutes.

From Kepler's third law, we know that 300-km-high satellite will have a period of 90.37 minutes. If the satellite orbital inclination is 60° and the target is at 35° latitude, the formula says that the satellite will see the signal for 12.2 minutes if it passes directly over the emitter.

9.2 HORIZON PLOT ON THE EARTH

This section shows how to generate an Earth surface contour of the horizon from a satellite. The discussion includes examples for a 300-km-high satellite with its SVP at 100° East longitude and 35° North latitude.

Consider the plane right triangle in Figure 9.12.

$$\text{Angle } C = a\cos(\text{Side } a / \text{Side } c)$$

where a is the radius of the Earth and c is the altitude of the satellite plus the radius of the Earth.

Thus, the geocentric angle (C) to the horizon from the SVP is:

$$C = \arccos\left[R_E / (h + R_E)\right]$$

For a 300-km-high satellite, C would be $\arccos[6{,}371/6{,}671] = 17.2°$.

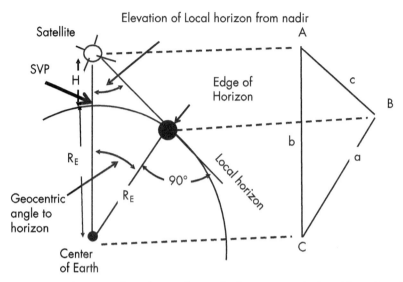

Figure 9.12 The range from the satellite to the horizon is calculated from this plane triangle.

Now consider the spherical triangle of Figure 9.13, formed by the SVP, the North Pole, and the edge point of the horizon trace on the Earth's surface.

- Side d is the same as angle C in Figure 9.12 (i.e., the geocentric angle from the SVP to the edge of the horizon).
- Side e is 90° minus the latitude of the SVP (i.e., 55°).
- Side f is 90° minus the latitude of the edge point of the horizon.
- Angle F is assigned as we are calculating the horizon edge points. It will be chosen every few degrees from 0° to 360°.
- Angle D is the difference in longitude between the SVP and the calculated edge point of the horizon.
- Angle E is at the point on the edge of the antenna footprint.

You can calculate a table of the latitude and longitude points for the edge of the horizon with the resolution you like (the computer will not mind generating and plotting a lot of points).

As an example, we will manually calculate only one point for our chosen satellite. Let's choose the horizon edge point 45° East of

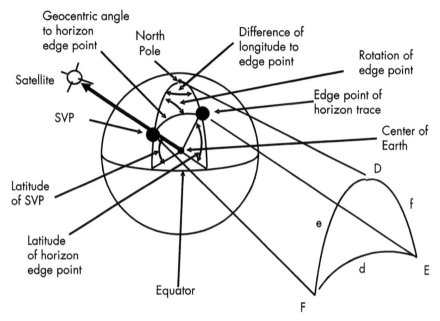

Figure 9.13 The location of point on the horizon relative to the satellite SVP is shown in this spherical triangle.

North. This makes angle F = 45°, side e is 55°, and side d is angle C from Figure 9.12 (i.e., 17.2°).

From the law of cosines for sides for spherical triangles:

$$\cos f = (\cos e)(\cos d) + (\sin e)(\sin d)(\cos F)$$

So

$$f = a\cos\left[(\cos e)(\cos d) + (\sin e)(\sin d)(\cos F)\right]$$

Let's choose the edge point 45° East of North. This makes angle F = 45°.

Side f is then:

$$a\cos\left[(\cos 55°)(\cos 17.2°) + (\sin 55°)(\sin 17.2°)(\cos 45°)\right]$$
$$= a\cos\left[(0.574)(0.955) + (0.819)(0.296)(0.707)\right] = a\cos[0.719] = 44.0°$$

which makes the latitude of the calculated horizon edge point 90° − 44.0° = 46.0° North.

Now, from the law of sines for spherical triangles,

$$\sin D / \sin d = \sin F / \sin f$$

so

$$D = a\sin\left[(\sin d)(\sin F)/\sin f\right] = a\sin\left[(\sin 17.2)(\sin 45°)/\sin 46°\right]$$
$$= a\sin[(0.295)(0.707)/0.719°] = a\sin 0.290 = 16.9°$$

Since the satellite SVP is at 100° East latitude, the longitude of the calculated horizon edge point is 100° + 16.9° = 116.9° East longitude.

The calculation of the rest of the horizon points and the plotting is left as an exercise for the reader (and the reader's computer). This exercise is repeated as angle F goes through 360°, with whatever resolution you require.

9.3 INTERCEPT OF THE EARTH SURFACE TARGET USING A NARROW-BEAM RECEIVING ANTENNA

In Section 9.1, we dealt with intercept from a satellite with a nondirectional antenna; now we are adding a directional antenna to the satellite. First, we will calculate the antenna look angles to any point on the Earth's surface that can be seen by the satellite.

9.3.1 Antenna Pointing

We can calculate the azimuth and elevation (from the nadir) required to direct the satellite antenna boresight at an Earth surface target at a specific latitude and longitude. The SVP is 100° East longitude and 30° North latitude, the same as stated for Sections 9.1 and 9.2, and the target is at 102° East longitude, 32° North latitude. Note that this is well within the area that the satellite can see. First consider the spherical triangle of Figure 9.14 formed by the SVP, the target location and the North Pole.

- h = 90° minus latitude of SVP (i.e., 60°);
- j = 90° minus latitude of target (i.e., 58°);

9.3 INTERCEPT OF THE EARTH SURFACE TARGET

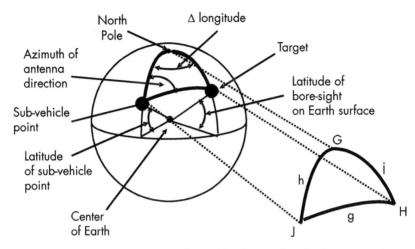

Figure 9.14 The spherical triangle formed by the North Pole, the SVP, and a target at a specified longitude and latitude allows calculation of the satellite antenna azimuth look angle to that target.

- g = geocentric angle from the SVP to the target;
- G = the difference in longitude between the SVP and the target (i.e., 2°);
- J = the azimuth to which the antenna must be directed.

From the law of cosines for sides:

$$\cos g = (\cos h)(\cos j + (\sin h)(\sin j)(\cos G))$$

So

$$g = a\cos[\cos 60°)(\cos 58°) + (\sin 60°)(\sin 58°)(\cos 2°)]$$
$$= a\cos[(0.500)(0.530) + (0.866)(.848)(.999)] = a\cos[0.738] = 2.6°$$

This is the geocentric angle between the satellite and the target. Then, from the law of sines for spherical triangles,

$$\sin J / \sin j = \sin G / \sin g$$

So

$$J = a\sin\bigl[(\sin G)(\sin j)/\sin g\bigr] = a\sin\bigl[(\sin 2°)(\sin 58°)/\sin 2.6°\bigr]$$
$$= a\sin\bigl[(0.035)(0.848)/0.045\bigr] = a\sin[.660] = 41.3°$$

This is the azimuth to the target from the satellite.

Now consider the plane triangle in Figure 9.15 formed by the satellite, the target, and the center of the Earth.

- $m = R_E$ + height of satellite;
- $k = R_E$;
- n = the propagation range from the satellite to the target;
- K = the elevation (from nadir) from the satellite to the target;
- N = the geocentric angle from the satellite to the target;
- M = the angle from the center of the Earth to the satellite as seen from the target.

Angle N in Figure 9.15 is side g from Figure 9.14 (i.e., 2.6°). Using the law of cosines for plane triangles:

$$n^2 = m^2 + k^2 - 2nk \cos N$$

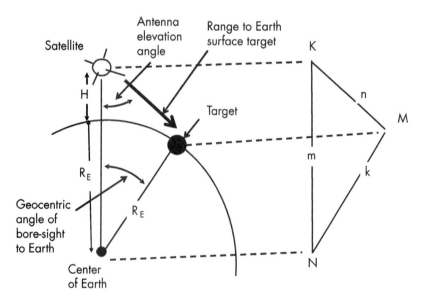

Figure 9.15 The elevation of the antenna (above the nadir) and the range from the satellite to a target on the Earth's surface can be calculated from this plane triangle.

9.3 INTERCEPT OF THE EARTH SURFACE TARGET

So

$$n = \text{sqrt}[m2 + k2 - 2mk\cos N]$$
$$= \text{sqrt}\left[6617^2 + 6371^2 - 2(6671)(6371)(\cos(2.6°))\right]$$
$$= \text{sqrt}[40589641 + 44502241 - 84914378] = \text{sqrt}[177504] = 421 \text{ km}$$

This is the range from the satellite to the target.
Then, from the law of sines for plane triangles,

$$\sin K/k = \sin N/n$$

So

$$K = a\sin[k * \sin N / n]$$
$$= a\sin\left[(6371)(\sin(2.6°))/421\right] = a\sin[(6371)(.045)/418] = 68.6°$$

Now we can find angle $M = 180° - 2.6° - 68.6° = 108.8°$.
This is the elevation (from the nadir) of the pointing angle.
The local horizon at the boresight point is 90° above the nadir, so the satellite is $108.8° - 90° = 18.8°$ above the horizon.

9.3.2 Intercept Link Equation

We will intercept a 4-GHz communication emitter at 105° East longitude, 35° North latitude in heavy rain. The ERP of the target emitter is 100W (50 dBm) and it has a very wide antenna pattern. The satellite has a 1-m parabolic antenna that is aimed at the target emitter with a pointing accuracy of 1°.

Antenna Pointing Angles

Using the procedure presented in Section 9.3.1, we will calculate the azimuth and elevation (from the nadir) required to direct the satellite antenna boresight at the target transmitter. First consider the spherical triangle of Figure 9.16 formed by the SVP, the target location, and the North Pole. Note that this is the same as Figure 9.14, but we have plugged in the values for the current problem.

- $h = 60°$;

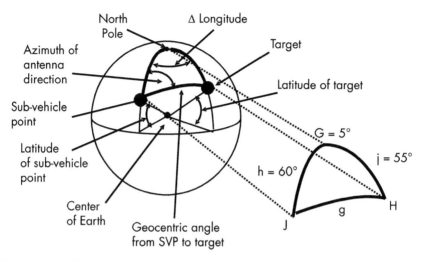

Figure 9.16 The spherical triangle formed by the North Pole, the SVP, and a target at a specified longitude and latitude allows the calculation of the satellite antenna azimuth look angle and geocentric angle to that target.

- $j = 55°$;
- $G = 5°$;
- g = geocentric angle from satellite to target;
- J = the azimuth to which the antenna must be directed.

From the law of cosines for sides:

$$\cos g = (\cos h)(\cos j) + (\sin h)(\sin j)(\cos G)$$

So

$$g = a\cos\left[(\cos 60°)(\cos 55°) + (\sin 60°)(\sin 55°)(\cos 5°)\right]$$
$$= a\cos\left[(0.500)(0.574) + (0.866)(0.819)(0.996)\right] = a\cos[0.993] = 6.8°$$

Then, from the law of sines for spherical triangles,

$$\sin J / \sin j = \sin G / \sin g$$

So

9.3 INTERCEPT OF THE EARTH SURFACE TARGET

$$J = a\sin\left[(\sin G)(\sin j)/\sin g\right] = a\sin\left[(\sin 5°)(\sin 55°)/\sin 6.8°\right]$$
$$= a\sin\left[(0.087)(0.819)/0.118\right] = a\sin[.603] = 37.1°$$

This is the azimuth to the target hostile transmitter from the satellite.

Now consider the plane triangle in Figure 9.17 formed by the satellite, the target, and the center of the Earth. This is Figure 9.15 with the values from the current problem plugged in.

- $m = 6{,}671$ km;
- $k = 6{,}371$ km;
- n = the propagation range from the satellite to the target;
- K = the elevation (from the nadir) from the satellite to the target;
- N = the geocentric angle from the satellite to the target;
- M = the angle from the center of the Earth to the satellite as seen from the target.

Angle N in Figure 9.17 is side g from Figure 9.16 (i.e., 6.8°).

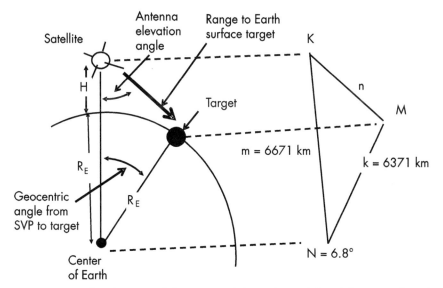

Figure 9.17 The elevation of the antenna (above the nadir) and the range from the satellite to a target on the Earth surface can be calculated from this plane triangle.

Using the law of cosines for plane triangles,

$$n^2 = m^2 + k^2 - 2nk\cos N$$

So

$$n = \text{sqrt}\left[m^2 + k^2 - 2mk\cos N\right]$$
$$= \text{sqrt}\left[6617^2 + 6371^2 - 2*6671*6371*\cos(6.8°)\right]$$
$$= \text{sqrt}[40589641 + 44502241 - 84406869] = \text{sqrt}[685013] = 828\ km$$

This is the range from the satellite to the target transmitter. Then, from the law of sines for plane triangles,

$$\sin K / k = \sin N / n$$

So

$$K = a\sin[k \sin N / n] = a\sin[6371)(\sin(6.8°)/828]$$
$$= a\sin[(6371)(.118)/828] = 65.6°$$

Now we can find angle $M = 180° - 6.8° - 65.6° = 107.8°$. This is the elevation (from the nadir) to the target transmitter from the satellite.

The local horizon at the boresight point is 90° above the nadir, so the satellite is $107.8° - 90° = 17.8°$ above the horizon from the target emitter. We will need this to calculate the rain loss.

9.3.3 Link Losses

The link losses from the target location can now be calculated. They include the spreading loss, the satellite antenna pointing error, atmospheric attenuation, and rain loss.

The spreading loss is found from the formula:

$$L = 32.44 + 20\log(F) + 20\log(d)$$

where L is the spreading loss in decibels, F is the emitter transmit frequency (4 GHz), and d is the link distance (n from above = 828 km).

$$L = 32.44 + 20\log(4000) + 20\log(828) = 32.44 + 72.04 + 58.36 = 162.8 \ dB$$

The atmospheric loss is a function of the elevation angle of the satellite above the target emitter's horizon, which we calculated to be 18.8°. The atmospheric loss can be determined from Figure 9.18. At 4 GHz and 18.8° elevation, this loss is about 0.1 dB.

The 3-dB beamwidth of the 1-m receiving antenna at 4 GHz can be calculated from the formula:

$$BW = \text{anti}\log\left[(86.8 - 20\log D - 20\log F)/20\right]$$

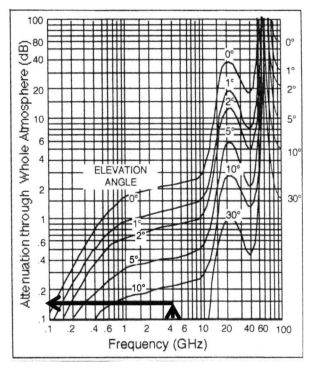

Figure 9.18 At 4 GHz, an elevation of 18.8 causes 0.15 dB of atmospheric attenuation.

where *BW* is the 3-dB beamwidth of a 55% efficient parabolic antenna, *D* is the diameter of the antenna in meters, and *F* is the frequency in megahertz.

$$BW = \text{anitlog}(86.8 - 0 - 72.0) = 6.3°$$

Now we can calculate the antenna misalignment loss from the formula in Section 5.2.

$$\Delta G = 12(\theta/\alpha)^2$$

where ΔG is the misalignment loss in decibels, θ is the antenna misalignment loss in degrees, and α is the 3-dB beamwidth of the antenna.

At 1° antenna misalignment,

$$\Delta G = 12(1/6.3)^2$$
$$\Delta G = 0.3 dB$$

The distance through the rain can be calculated from Figure 9.19. The height of the 0° isotherm is 3 km (at 35° latitude as shown in Figure 9.5). The distance through the rain is then:

$$\text{Distance} = 3\text{km}/\sin(18.8°) = 3\text{km}/0.322 = 9.3\text{km}$$

From Figure 9.20, the loss in heavy rain at 4 GHz is about 0.02 dB/km. Multiplied by 9.8 km, this makes the heavy rain loss = 0.2 dB.

Satellite to Target Link Loss

The total link loss is thus:
Spreading loss + Antenna misalignment loss + Atmospheric loss + Rain loss
The link loss = 162.8 dB + 0.3 dB + 0.1 dB + 0.2 dB = 163.4 dB.
The gain of the 1-m satellite antenna can be calculated from the formula:

$$G = -42.2 + 20\log D + 20\log F$$

9.3 INTERCEPT OF THE EARTH SURFACE TARGET

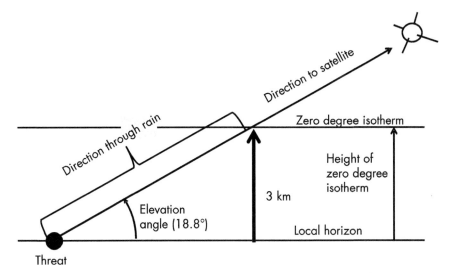

Figure 9.19 The distance over which rain loss applies is the part of the link below the 0° isotherm, which is 3 km, as shown in Figure 9.5.

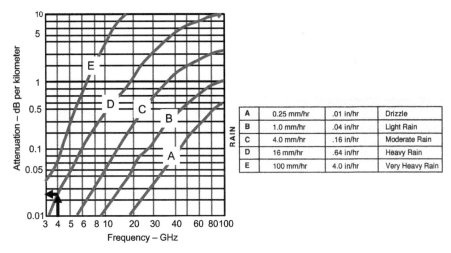

Figure 9.20 The loss from heavy rain at 4 GHz is 0.02 dB/km.

where G is the gain of a 55% efficient parabolic antenna, D is the diameter of the antenna in meters, and F is the frequency in megahertz.

$$G = -42.2 + 0 + 72 = 29.8 \text{ dB}$$

The received power in the satellite receiver is:

$$P_R = ERP - L + G_R$$

where P_R is the received power, L is the total link loss, and G_R is the receiving antenna gain.

$$P_R = 50 \text{ dBm} - 163.4 \text{ dB} + 29.8 \text{ dB} = -83.6 \text{ dBm}$$

So this means that the satellite receiver must have sensitivity of −83.6 dBm to successfully intercept this target signal in this situation.

Receiver Sensitivity

First, we will calculate the receiver sensitivity required to intercept the signal if the satellite is on the horizon. Then we will determine how long the satellite can see the signal if the orbit passes directly over the emitter. Finally, we will determine the sensitivity that would be required if the satellite were in a synchronous orbit.

9.3.4 Intercept from the Horizon

The satellite is in a circular orbit 300 km above the Earth. We want to intercept a 4-GHz target emitter from its horizon in heavy rain. The ERP of target emitter is 100W (50 dBm) and it has a very wide antenna pattern. (We assume that it covers from horizon to horizon.) The satellite has a 1-m parabolic antenna that provides 29.8-dBi gain. It is aimed at the target emitter with a pointing accuracy of 1° (which causes a misalignment loss of 0.3 dB).

We will be using equations that have been presented in earlier chapters to provide the answers for this problem.

The plane right triangle of Figure 9.21 shows the geometry to the horizon. We want to determine the distance from the satellite to the horizon, which is c in the right-hand triangle. Side b is the semi-major axis of the satellite (6,671 km), and side a is the radius of the Earth (6,371 km). Since angle B is 90°, c is found from the formula:

$$\text{Side } c = \text{sqrt}\left[b^2 - a^2\right] = \text{sqrt}\left[(6671)^2 - (6371)^2\right] = 1978 \text{ km}$$

This means that the LOS loss is:

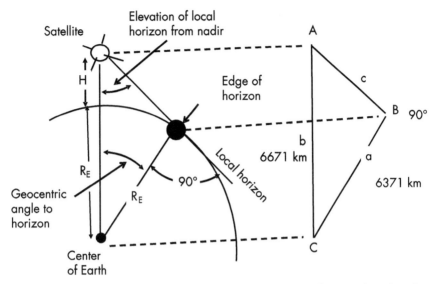

Figure 9.21 The range from the satellite to the horizon is calculated from this plane triangle.

$$LOS = 32.4 + 20\log(F) + 20\log(d)$$

where LOS is the LOS loss in decibels, F is the frequency in megahertz, and d is the link distance in kilometers.

$$LOSS = 32.4 + 20\log(4{,}000) + 20\log(1{,}978) = 170.4 \text{ dB}$$

In Section 9.3.3, we determined that the antenna-pointing accuracy loss is 0.3 dB.

From Figure 9.22, we can see that the atmospheric loss through the whole atmosphere at 4 GHz and 0° elevation is 2.2 dB.

We know that the rain loss is a function of the distance from the 0° isotherm to the ground emitter along the signal transmission path.

Figure 9.23 shows the geometry of the rain distance if we use the same height for the 0° isotherm from above (i.e., 3 km). The point at which the signal path crosses the 0° isotherm, the target, and the center of the Earth form a right triangle. The rain distance is then defined by:

$$\text{Rain distance} = \text{sqrt}\left[(6{,}374)^2 - (6{,}371)^2\right] = 196 \text{ km}$$

178 INTERCEPT FROM SPACE

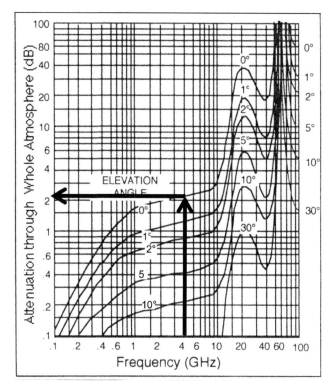

Figure 9.22 At 4 GHz, an elevation of 0 causes 2 dB of atmospheric attenuation.

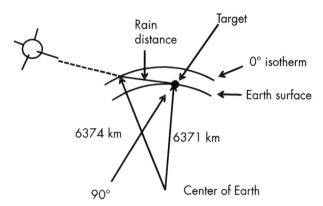

Figure 9.23 The distance through the rain is from the ground to the point at which the signal path to the satellite passes the 0° isotherm.

9.3 INTERCEPT OF THE EARTH SURFACE TARGET

We established from Figure 9.20 that the heavy rain loss at 4 GHz is 0.02 dB/km. This would make the rain loss 196 × 0.02 dB = 4 dB. However, heavy rain over 196 km would cause a major flood, so we will assume that the rain is concentrated in heavy rain cells over 20% of the rain distance, so we will use 0.8 dB as our budgeted rain loss.

Now we can calculate the total link loss:

Total link loss = LOS loss + antenna misalignment loss
+ atmospheric loss + rain loss
= 170.4 dB + 0.3 dB + 2.2 dB + 0.8 dB = 173.7 dB

Now we can calculate the receiver system sensitivity required to intercept the target signal at the satellite's horizon.

Received power = Target signal ERP − Total link loss
+ Receiving antenna gain
= 50 dBm − 173.7 + 29.8 dBi = −93.9 dBm

We have not discussed the modulation of the target signal, so this sensitivity may or may not be achievable. If not, we might have to do something like increase the size of the antenna.

Duration of the Intercept

By Kepler's third law, the relationship between the period and the semi-major axis of a satellite is:

$$a^3 / P^2 / C = 36{,}355{,}285 \text{ km}^3 / \text{min}^2$$

so P squared is a cubed divided by 36,355,285, which equals 8,165.9 km squared.

The semi-major axis of a 300-km-high satellite is 6,671 km, so its period must be 90.365 minutes.

Here is the ugly equation for horizon to horizon observation time in Section 8.3.

$$T_{TOTAL} = P\bigl[2\arccos(R_E/a)\bigr]\bigl[1+\cos(i)\cos(lat)\bigr]$$

To plug in some numbers, let's use the threat location latitude (35°) and let the orbital inclination = 60°.

The time that the satellite can "see" the target in one direct overhead pass is, therefore, 12.2 minutes.

9.4 INTERCEPT FROM THE SYNCHRONOUS SATELLITE

Figure 9.24 shows the distance from the satellite to an intercept target on the Earth. If the satellite is on the horizon from the target, the range is 41,759 km, and if the satellite is directly over the target, the range is 35,873 km. As we will calculate, the losses from these ranges are high, but the synchronous satellite gives the advantage that it will receive the signal continuously.

9.4.1 With the Satellite on the Horizon

The ERP of the target transmitter is 100W (50 dBm). The LOS loss to the satellite on the horizon is:

$$LOS\ loss = 32.4 + 20\log(4,000) + 20\log(41,682) = 196.8\ dB$$

The atmospheric and rain loss are the same as for the 300-km-high satellite at the target's horizon (2.2 and 0.8 dB, respectively).

Let's assume that the satellite intercept antenna is 1m in diameter and the signal to be intercepted is at 4 GHz. Using the gain formula presented in Section 5.1.

$$G = -42.2 + 20\log(D) + 20\log(F)$$

where G is antenna gain (in decibels), D is the reflector diameter (in meters), F is the frequency (in megahertz), and the antenna efficiency is 55%.

The gain of the antenna is 29.8 dBi, which we suspect is not enough. However, if we increase the antenna diameter to 5m, the gain would be 43.8 dBi at 4 GHz.

Using the beamwidth formula from Section 5.1, the 3-dB beamwidth of the antenna is given by:

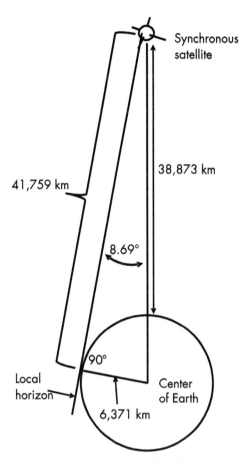

Figure 9.24 The range to a synchronous satellite that is on the horizon is 41,759 km and from directly overhead it is 35,873 km.

$$20\log(BW) = 86.8 - 20\log(D) - 20\log(F)$$

where BW is the 3-dB beamwidth in degrees, D is the reflector diameter in meters, and F is frequency in megahertz.

At 5-m diameter and 4-GHz frequency, the 3-dB beamwidth is:

$$\text{antilog}\big[(86.8 - 20\log(5) - 20\log(4000))/20\big] = 1.1°$$

Using the formula from Section 5.2.

$$\Delta G = 12(\theta/\alpha)^2$$

where G is the reduction from boresight gain in decibels, α is the 3-dB beamwidth in degrees, and θ is the offset angle in degrees.

If we aim the satellite antenna with 1/4° accuracy, the misalignment error would cause:

$$12(0.25/1.1)^2 = 0.6 \text{ dB loss}$$

The total loss is 196.8 dB + 2.2 dB + 0.8 dB + 0.6 dB = 200.4 dB. The received signal strength is then 50 dBm − 200.4 dB + 43.8 dBi = −106.8 dBm.

If the satellite intercept receiver has a noise figure of 2 dB and the required bandwidth to receive the target signal is 1 GHz, the minimum discernable signal level of the receiver would be −114 dBm + 2 dB = −112 dBm, so the target signal would be received with only 5.2-dB signal-to-noise ratio.

If you needed a better signal quality (let's say, 15-dB signal-to-noise ratio), you would need 9.8-dB more antenna gain (53.6 dBi). This would require a 15.4-m antenna.

9.4.2 With the Satellite Directly Overhead

If the satellite SVP is on the target, the distance from the satellite to the target is 35,795 km, so the LOS loss is:

$$\begin{aligned}\text{LOS loss} &= 32.4 + 20\log(4,000) + 20\log(35,795) \\ &= 32.4 + 72 + 91.1 = 195.5 \text{ dB}\end{aligned}$$

However, the atmospheric and rain losses are zero because of the vertical elevation of the satellite from the target transmitter. So the total loss is 195.5 dB. The received signal strength is then 50 dBm − 195.5 dB + 43.8 dBi = −101.7 dBm.

With the same receiver and signal bandwidth (2-dB noise figure and 1 MHz) but the satellite directly over the target transmitter, the received signal strength would be −101.7 dBm, so the received signal-to-noise ratio would be 10.5 dB.

If you need 15-dB signal-to-noise ratio, the intercept antenna diameter would need to provide 4.5-dB more gain. Thus, the diameter must be increased to 10m.

10

JAMMING FROM SPACE

10.1 JAMMING OF A GROUND SIGNAL FROM A SATELLITE

In Chapter 9, we discussed the intercept of hostile ground signals from space. Now we will discuss the jamming of hostile ground signals from space. We will consider both communication jamming and radar jamming. Jamming from space requires that we consider the geometric dictates of the orbit from which the jamming takes place.

Note that the jamming formulas used in this chapter are defined in Appendix A.

Another important note about this chapter is that the satellite and orbit specifications and the target specifications called out in problems are not intended to repeat those of any real-world satellites or targets. The numbers are just chosen to be reasonable, so we can talk about the equations and the results they would produce. You can apply the equations to any real-world problems that you need to analyze in the future by plugging in the real-world specifications.

10.2 JAMMING FROM A SATELLITE

The effectiveness of a jammer requires that it produce the proper waveform, that it has adequate ERP, and that it is close enough to the jammed receiver to keep the propagation losses low enough. Satellites, by their nature, are far away from what is happening on the

Earth, so jamming from space is not easy. That said, space is a well-established EW battlespace, and technologies are advancing, so it rates serious consideration.

For convenience, we will use the same orbit that we used in Chapter 9, Section 9.3: a 300-km-high circular orbit with an inclination of 60°. We will place the SVP at 30° North latitude and 100° East longitude and will jam a target at 32° North latitude and 102° East longitude. We know that the period of the orbit is 90.364 minutes from Kepler's third law and that, if it passes directly over the target, it can see the target for 12.2 minutes (because we calculated that in Chapter 9).

The satellite payload is a 100-W jammer with a 2-m dish antenna.

First consider the spherical triangle of Figure 10.1 formed by the SVP, the target location, and the North Pole:

- c = 90° – latitude of SVP (i.e., 60°);
- b = 90° – latitude of target (i.e., 58°);
- a = geocentric angle from the SVP to the target;
- A = the difference in longitude between the SVP and the target (i.e., 2°);
- B = the azimuth to which the antenna must be directed.

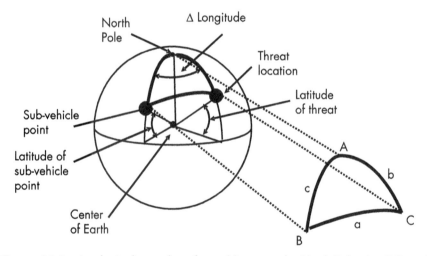

Figure 10.1 A spherical triangle is formed between the North Pole, the SVP, and the threat location.

10.2 JAMMING FROM A SATELLITE

From the law of cosines for sides:

$$\cos a = (\cos c)(\cos b) + (\sin c)(\sin b)(\cos G)$$

So

$$a = a\cos[(\cos 60°)(\cos 58°) + (\sin 60°)(\sin 58°)(\cos 2°)]$$
$$= a\cos[(0.500)(0.530) + (0.866)(0.848)(0.999)] = a\cos[0.738] = 2.6°$$

This is the geocentric angle between the satellite and the target.

Now consider the plane triangle in Figure 10.2 formed by the satellite, the target, and the center of the Earth.

- $f = R_E$ + height of satellite;
- $e = R_E$;
- g = the propagation range from the satellite to the target;
- G = the elevation (from the nadir) from the satellite to the target;
- F = the geocentric angle from the satellite to the target;

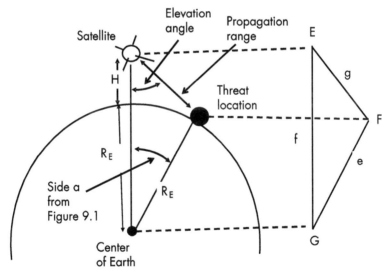

Figure 10.2 The elevation from nadir and range to a threat from a satellite can be determined from the plane triangle defined by the satellite, the threat, and the center of the Earth.

- G = the angle from the center of the Earth to the satellite as seen from the target.

Angle N in Figure 10.2 is side a from Figure 10.1 (i.e., 2.6°). Using the law of cosines for plane triangles:

$$g^2 = f^2 + e^2 - 2fe\cos G$$

So

$$\begin{aligned}g &= \operatorname{sqrt}\left[f^2 + e^2 - 2fe\ \cos G\right] \\ &= \operatorname{sqrt}[6{,}617^2 + 6{,}371^2 - 2(6{,}671)(6{,}371)(\cos(2.6°))] \\ &= \operatorname{sqrt}[40{,}589{,}641 + 44{,}502{,}241 - 84{,}914{,}378] \\ &= \operatorname{sqrt}[177{,}504] = 421\ km\end{aligned}$$

This is the range from the satellite to the target as shown in Figure 10.3.

10.3 JAMMING OF A COMMUNICATIONS NETWORK

10.3.1 The Network

Consider the communications jamming engagement shown in Figure 10.4. Although this figure shows one desired signal transmitter and one target receiver, the actual target is a network of hostile transceivers as shown in Figure 10.5 operating at 400 MHz with whip antennas. One transceiver is transmitting with an ERP of 10W and all of the others in the network are receiving. The average distance to each receiving station is 5 km. The SVP of the satellite is 30° North latitude and 100° East longitude. The jammer transmitter output power is 100W and its transmitting antenna is 2m in diameter. It is aimed with one degree accuracy. The target communications network is located at 32° North latitude and 102° East longitude.

10.3.2 Link Equations

The 3-dB beamwidth and boresight gain of the jamming antenna and the antenna misalignment loss are determined by formulas from Chapter 5:

10.3 JAMMING OF A COMMUNICATIONS NETWORK 187

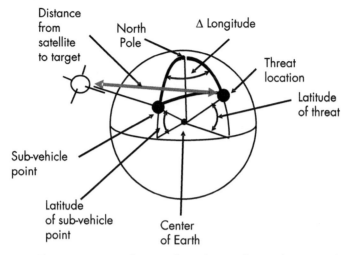

Figure 10.3 The propagation distance from the satellite to the target depends on the orbital geometry.

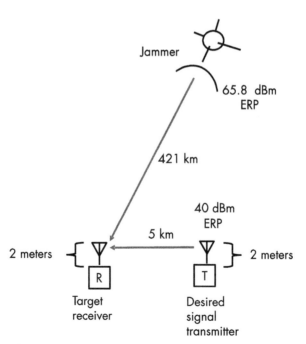

Figure 10.4 The jamming geometry and the ERPs of the desired signal and jamming transmitters determine the J/S.

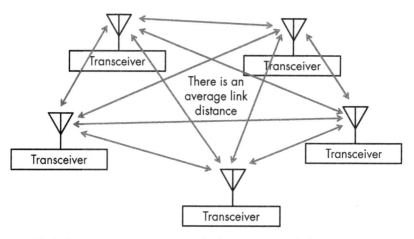

Figure 10.5 In a communication network, there are many links; one station is transmitting but all of the others are receiving.

$$\text{Beamwidth:} BW = \text{antilog}\left[(86.8 - 20\log D - 20\log F)/20\right]$$

where BW is the 3-dB beamwidth in degrees, D is the dish diameter in meters, and F is the frequency in megahertz.

$$BW = \text{antilog}\left[(86.8 - 6.0 - 52.0)/20\right] = 27.5°$$
$$\text{Gain: } G = -42.2 + 20\log D + 20\log F$$

where G is the boresight gain in dBi.

$$G = -42.2 + 6.0 + 52.0 = 15.8 \text{ dBi}$$
$$\text{Misalignment loss:} \Delta G = 12(\theta/\alpha)^2$$

where ΔG is the loss in decibels from antenna misalignment, θ is the antenna misalignment, and α is the antenna 3-dB beamwidth.

$\Delta G = 12(1/15.8)^2 = 0.05$ dB (which we can ignore in our calculations)

The equation for the J/S in communications jamming if the target receiver has a whip antenna or some other antenna with a 360° horizontal antenna pattern is:

$$J/S = ERP_J - ERP_S - LOSS_J + LOSS_S$$

10.3 JAMMING OF A COMMUNICATIONS NETWORK

where J/S is the jamming-to-signal ratio in decibels, ERP_J is the effective radiated power of the jammer, ERP_S is the effective radiated power of the desired signal transmitter, $LOSS_J$ is the loss from the jammer to the target receiver, and $LOSS_S$ is the loss from the desired signal transmitter to the target receiver.

Now we will determine the propagation modes and losses from the desired signal transmitter and the jammer to the target receiver using the formulas presented in Chapter 5.

We will assume that there is clear LOS between all of the members of the hostile network that we are jamming. Because the frequency is ultrahigh frequency (UHF) and the satellite is at a high angle as viewed from the target receiver, we can ignore atmospheric and rain losses (they are very small).

The formula for the Fresnel zone distance is

$$FZ = (h_T \times h_R \times F) / 24{,}000$$

where FZ is the Fresnel zone distance in kilometers, h_T is the height of the transmit antenna in meters, h_R is the height of the receive antenna in meters, and F is the transmission frequency in megahertz.

For the desired signal link, the two antenna heights are 2m and the frequency is 400 MHz, so

$$FZ = (2 \times 2 \times 400) / 24{,}000 = 67 \; meters$$

Since FZ is less than the link distance, the propagation mode is two-ray, so the propagation loss is:

$$LOSS_S = 120 + 40\log(d) - 20\log(h_T) - 20\log(h_R)$$

where d is the link distance in kilometers.
The desired link loss is thus:

$$120 + 40\log(5) - 20\log(2) - 20\log(2)$$
$$= 120 + 28.0 - 6.0 - 6.0 = 136 \; dB$$

The jamming link propagation mode is LOS (also called free space, which it literally is in this case). The loss from antenna misalignment is extremely small, so the total loss from the jammer to the target receiver is:

$$LOSS_J = 32.4 + 20\log(d) + 20\log(F) = 32.4 + 20\log(421) + 20\log(400)$$
$$= 32.4 + 52.5 + 52.0 = 136.9 \text{ dB}$$

The ERP of the desired signal transmitter is 10W (40 dBm). The ERP of the jammer is 100W (50 dBm) increased by the antenna gain (15.8 dBi) = 65.8 dBm.

10.3.3 J/S

The J/S is described in Appendix A; the formula is:

$$J/S = ERP_J - ERP_S - LOSS_J + LOSS_S$$

For this problem:

$$J/S = 65.8 \text{ dBm} - 40 \text{ dBm} - 136.9 \text{ dB} + 136 \text{ dB} = 24.9 \text{ dB}$$

This is a very respectable job of jamming (10 dB would usually be enough). However, it must be noted that one satellite can only jam that network for up to 12.2 minutes, so an array of satellites would be required to perform continuous jamming.

Note that if the satellite had been at the horizon from the jammed network (i.e., 1,978 km away), the achieved J/S would have been reduced to 10.5 dB (including the effect of 1 dB of atmospheric loss to the jamming signal).

10.4 JAMMING A MICROWAVE DIGITAL DATA LINK

As shown in Figure 10.6, there is an unmanned aerial vehicle (UAV) that is passing data over a link to its control station. The satellite is jamming that link. The satellite is in the same orbit we used in Section 10.3, and the SVP and target location are the same.

The target receiver being jammed is the receiver of a 5-GHz downlink from the UAV. The jammed link has a 10-W transmitter with a circularly polarized phased array antenna that has 10-dBi gain. The hostile UAV has ERP of 10W (40 dBm) at 5 MHz. For this problem,

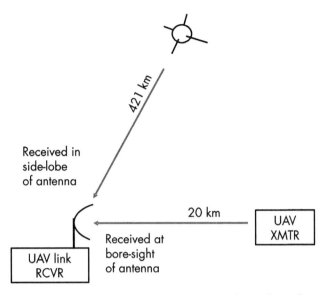

Figure 10.6 The directional receiving antenna is aimed at a desired signal transmitter. The satellite is in a sidelobe of the receiving antenna.

the UAV is 20 km from its control station, which includes the targeted link receiver. The targeted receiving station has a 1-m parabolic antenna that has sidelobes with gain 20 dB below its boresight gain.

Since the orbital and jamming geometries are the same as those of Section 10.1, we will not repeat those calculations. The distance from the satellite to the target receiver is still 421 km.

First, calculate the loss from the UAV to its ground station. Since this is a microwave link, it can be assumed to have LOS propagation:

$$L = 32.4 + 20\log(d) + \text{to}\log(F)$$

where L is the link loss in decibels, d is the link distance in kilometers, and F is the link frequency in megahertz.

$$\text{Loss} = 32.4 + 20\log(20) + 20\log(5,000) = 32.4 + 26 + 74 = 132.4 \text{ dB}$$

The atmospheric and rain losses will be very small for the UAV link, so we will ignore them.

The gain of the UAV link receiving antenna at its boresight is

Gain: $G = -42.2 + 20\log D + 20\log F$

where G is the boresight gain in dBi, D is the diameter of the antenna in meters, and F is the link operating frequency in megahertz.

$$G = -42.2 + 0 + 74 = 31.8 \text{ dBi}$$

Its average sidelobe gain is 20 less or 11.8 dBi.
The UAV link transmitter ERP is 10W (40 dBm).
Now consider the jamming link. It is at 5 GHz over a distance of 421 km and has LOS propagation, so the link loss is:

$$\text{Loss} = 32.4 + 20\log(421) + 20\log(5{,}000) = 32.4 + 2.5 + 74 = 158.9$$

Because the satellite is at a very high elevation angle to the target receiver, the atmospheric and rain losses will be very low, so we will ignore them for this problem.

The gain of the satellite's 3-m transmit antenna is:

$$\text{Gain:} G = -42.2 + 20\log D + 20\log F$$

where G is the boresight gain in dBi, D is the diameter of the antenna in meters, and F is the link operating frequency in megahertz.

$$\begin{aligned}\text{Gain:} G &= -42.2 + 20\log D + 20\log F \\ &= -42.2 + 20\log(3) + 20\log(5000) = -42.2 + 9.5 + 74 = 41.3 \text{ dBi}\end{aligned}$$

The beamwidth of the satellite antenna can be found from the formula:

$$\alpha = \text{antilog}\left[(86.8 - 20\log D - 20\log F)/20\right]$$

where D is the antenna diameter in meters and F is the frequency in megahertz.

$$\begin{aligned}\alpha &= \text{antilog}\left(\left[86.8 - 20\log(3) - 20\log(5.000)\right)/20\right] \\ &= \text{antilog}\left[(86.8 - 9.5 - 74)/20\right] = \text{antilog}\left[3.3/20\right] = 1.5°\end{aligned}$$

10.4 JAMMING A MICROWAVE DIGITAL DATA LINK

If the satellite antenna has a pointing accuracy of 1°, the reduction of the gain toward the target is given by the formula

$$\Delta G = 12(\Theta / \alpha)^2$$

where Θ is the antenna misalignment angle and α is the antenna's 3-dB beamwidth.

$$\Delta G = 12(1/1.5)^2 = 5.3 \text{ dB}$$

So the effective jammer boresight gain is 41.3 dBi – 5.3 dB = 36 dBi. This makes the ERP of the jammer toward the target 86 dBm.

The satellite payload jammer payload has a transmitter power of 100W (+50 dBm) with 41.3-dBi antenna gain, so the jammer ERP is 91.3 dBm.

Since the target receiver has a directional antenna, the J/S is by the formula:

$$J/S = ERP_J - ERP_S - L_J + L_S + G_{RJ} - G_R$$

where J/S is the ratio of the jammer power to the desired signal power at the input to the receiver being jammed (in decibels), ERP_J is the effective radiated power of the jammer (in dBm), ERP_S is the effective radiated power of the desired signal transmitter in dBm), L_J is the propagation loss from the jammer to the receiver (in decibels), L_S is the propagation loss from the desired signal transmitter to the receiver (in decibels), G_{RJ} is the receiving antenna gain in the direction of the jammer (in decibels), and G_R is the receiving antenna gain in the direction of the desired signal transmitter (in decibels).

$$J/S = 86 \text{ dBm} - 40\text{dB} - 158.9 \text{ dB}$$
$$+132.4 \text{ dB} + 11.3 \text{ dBi} - 31.3 \text{ dBi} = -0.5 \text{ dB}$$

Since the UAV link is digitally modulated, this is pretty close to the 0-dB J/S required to give effective jamming when the UAV is 20 km from its ground station.

10.5 JAMMING OF A GROUND RADAR FROM SPACE

10.5.1 Radar Jamming from a Satellite

The distance from the satellite to the radar being jammed depends on the location of the jamming satellite, the jammed radar, and the radar's target. For convenience, we will assume that the jamming satellite orbit and the satellite's location in that orbit are the same as they were for the communication jamming problem of Section 10.3, as shown in Figure 10.7. It is a circular orbit at 300-km altitude with an inclination of 60°. The SVP of the satellite is at 30° North latitude and 100° East longitude.

The radar that we are jamming is located on the Earth at 32° North latitude and 102° East longitude. The period of the satellite is 90.364 minutes. In the above communication jamming problem, we have calculated the distance from the satellite to the jammed radar to be 421 km. The time that the jammer is over the horizon to the radar that it is jamming (if it passes directly over the target) will be 12.2 minutes.

Placing the satellite where it was in Section 9.3 (SVP at 30° North latitude and 100° East longitude), we calculated the distance from the satellite to the jammed radar to be 421 km.

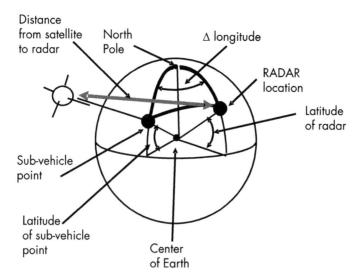

Figure 10.7 The propagation distance from the satellite to the target radar depends on the orbital geometry.

10.5.2 The Jammed Radar and Its Target

Now consider the radar we are attempting to jam, as shown in Figure 10.8. The range from the radar to its target is 15 km and the target radar cross-section is 10 m². We will set the radar's frequency at 5 GHz, its tube power at 100 kW, and its antenna diameter at 3m; 100 kW is 80 dBm. The radar antenna boresight gain is found from the formula:

$$G = -42.2 + 20\log D + 20\log F$$

where G is the boresight gain of the antenna in decibels, D is the diameter of the antenna in meters, and F is the operating frequency in megahertz.

The gain of the radar antenna is calculated to be

$$G = -42.2 + 9.5 + 74 = 41.3 \text{ dB}$$

Let's assume that the average sidelobes of the radar's antenna are 20 dB below the main beam boresight gain. Thus, the sidelobe gain is assumed to be 21.3 dB.

The ERP of the radar is:

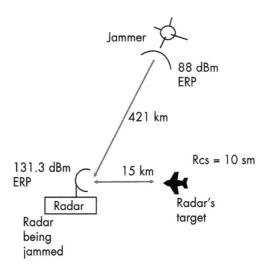

Figure 10.8 The jamming geometry and the ERPs of the desired signal and jamming transmitters determine the J/S.

$$ERP = \text{Tube power (in dBm)} + \text{Boresight gain in dB}$$
$$= 80 \text{ dBm} + 41.3 \text{ dB} = 121.3 \text{ dBm}$$

One more consideration is that the radar is aimed at its target, not at the satellite. Thus, the satellite must jam into the sidelobes of the radar antenna.

For this problem, let the radar cross-section of the radar's target be 10 m².

10.5.3 The Jammer

At 5 GHz, this 2-m antenna would have 37.8-dB gain (using the formula employed above for the target radar).

$$(-42.2 + 6 + 74) = 37.8 \text{ dB}$$

If the jammer transmitter has an output power of 100W, the ERP of the jammer will be:

$$50 \text{ dBm} + 37.8 \text{ dB} = 87.8 \text{ dBm}$$

The beamwidth of the jamming antenna can be found from the formula:

$$\alpha = \text{antilog}\left[(86.8 - 20 \log D - 20 \log F)/20\right]$$

where D is the antenna diameter in meters and F is the frequency in megahertz.

$$\alpha = \text{antilog}\left[(86.8 - 20\log(2) - 20\log(5,000))/20\right]$$
$$= \text{antilog}\left[(86.8 - 6 - 74)/20\right] = \text{antilog}[6.8/20] = 2.2°$$

If the jammer antenna is aimed to 1° accuracy, the reduction in gain is calculated from

$$\Delta G = 12(\Theta/\alpha)^2$$
$$= 12(1/2.2)^2 = 2.5°$$

10.5 JAMMING OF A GROUND RADAR FROM SPACE

So the effective ERP of the jammer toward the target radar is:

$$87.8 \text{ dBm} - 2.5 \text{ dB} = 85.3 \text{ dBm}$$

10.5.4 The Jamming Equation

Figure 10.4 shows the jamming geometry.

Using the formula for remote jamming from Appendix A, the signal-to-signal ratio (SNR) for remote radar jamming is given by:

$$J/S = 71 + ERP_J - ERP_S + 40\log R_T - 20\log R_J + G_S - G_R - 10\log RCS$$

where J/S is the jamming-to-signal ratio in decibels, ERP_J is the ERP of the jammer in dBm, ERP_S is the ERP of the radar in dBm, R_T is the range from the radar to its target in kilometers, R_J is the range from the jammer to the radar in kilometers, G_S is the sidelobe gain of the radar antenna in decibels, G_M is the main beam boresight gain of the radar antenna in decibels, and RCS is the radar cross-section of the radar's target in square meters.

Plugging in our given values, the J/S is:

$$J/S = 71 + 85.3 \text{ dBm} - 121.3 \text{ dBm} + 40\log(15) - 20\log(421)$$
$$+21.3 \text{ dB} - 41.3 \text{ dB} - 10$$
$$= 71 + 85.3 - 121.3 + 47 - 52.5 + 21.3 - 41.3 - 10 = -0.5 \text{ dB}$$

10.5.5 The Jamming Adequacy

The −0.5-dB J/S is not close to an adequate job of jamming. For adequate jamming, the J/S should probably be at least 10 dB. This is what would be required to protect against noise jamming, and it could only be continued for 12.2 minutes with a single satellite. Now let's look a few changes in the problem that would improve the quality of the jamming.

10.5.6 Protecting an Asset with Low Radar Cross-Section

If the radar's target has a 10^{-4}-m RCS (i.e., a stealth aircraft), the satellite jammer would provide a very comfortable 49.5-dB J/S.

$$(J/S = 71 + 85.3 - 121.3 + 47 - 52.5 + 21.3 - 41.3 + 40 = 49.5 \text{ dB})$$

This means that the jamming satellite could be in a much higher orbit; it could provide 10 dB from a range of 37,584 km, which is above the synchronous altitude.

10.5.7 Duration of Jamming

Remember that the jammer in a 300-km satellite is over the horizon to the jammed radar for only 12.2 minutes. This may be adequate for some tactical situation, but continuous jamming of the radar described in Section 10.5.2 would require several satellites. What if we use a synchronous satellite? The problem is that a synchronous satellite would be about 37,000 km from the jammed radar.

This will reduce the jamming signal (hence, the J/S) by about 40 dB relative to jamming from a low Earth satellite. The radar would have a hard time even detecting the presence of the jammer. Sorry, technical books do not always have to have happy endings (unless there is a huge technology breakthrough).

A

FORMULAS FROM SIGNAL INTELLIGENCE AND EW

The following formulas used in space EW come from signal intelligence and EW textbooks. This is not intended to be a complete coverage of these subjects; it just presents and explains the formulas used in this text.

Signal intelligence (SIGINT) is intelligence derived from received signals. Two subdivisions of this discipline are COMINT and ELINT. COMINT is intelligence derived from the reception and analysis of communications signals. ELINT is intelligence derived from the reception and analysis of noncommunications signals such as radars. SIGINT is often employed using satellites.

EW also has a component that employs the reception and analysis of hostile signals. This is called electronic support (ES). SIGINT and ES are close and often overlap. The most common division between the two (from a purely technical point of view) is that SIGINT finds out what the enemy has and ES finds out what assets the enemy is using against you right now. The timelines for SIGINT are fairly long; I was involved with a satellite that collected a signal from a new type of radar. It took months, and the satellite only got a few seconds of signal, but everyone was happy. However, EW must find out what enemy assets are being used right now before the enemy can use them to kill you. The necessary timelines are single-digit numbers of seconds.

EW also has a subfield called electronic attack (EA). EA includes all of the ways that action is taken to reduce the effectiveness of hostile military assets by electronic means. This includes attacks on all types of enemy radars and communications. Methods include jamming, chaff, flares, antiradiation weapons, and high-power radiation. From satellites, this is primarily jamming.

Another part of EW is electronic protection (EP). This involves features of communications systems and radars designed to prevent the successful application of EA.

Satellites can be a good choice for the implementation of both SIGINT and EW depending on the specific situation and activity.

A.1 INTERCEPT FORMULAS

Figure A.1 shows an intercept link. The important parameters of this link are the hostile transmitter power, the transmit antenna gain in the direction of the intercept receiver, the distance to the intercept receiver (in this case, perhaps to a satellite), and the antenna configuration of the intercept site.

A.1.1 Successful Intercept

A successful intercept occurs when a receiver in the satellite receives a hostile signal with enough quality to allow recovery of the information carried by the signal. For communication signals, this could be

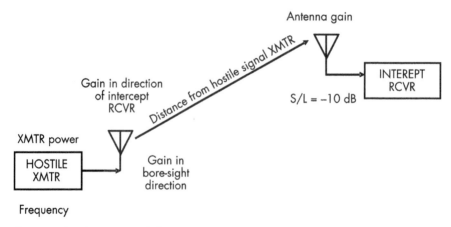

Figure A.1 An intercept link.

from either the externals or the internals of the signal. The externals include the transmission frequency, the modulation, or the type of encryption employed or the timing of the signal. It could also be the location of the transmitter. The internals of a communication signal is the information that it carries in its modulation. This can be voice, imagery, or data.

A.1.2 The Intercept Link Equation

In order to intercept a signal, the received signal strength must be high enough to exceed the sensitivity level of the receiver. The sensitivity is the weakest signal that the receiver can receive and still extract the required information from the signal.

The intercept link equation is:

$$P_R = ERP - Loss + G_R$$

where P_R is the power out of the receiving antenna in dBm, *ERP* is the effective radiated power of the hostile signal to be intercepted, *Loss* is the sum of all of the losses between the hostile transmitter's antenna and the intercept receiver's antenna (in decibels), and G_R is the receiving antenna gain (in decibels).

A.1.3 Received Signal Quality

Received signal quality is normally stated as the signal-to-noise ratio. This is the received signal level at the output of the receiving antenna divided by the noise level at same point in the receiver system.

The minimum discernable signal (MDS) is the received signal level when the signal power is equal to the noise power.

A.2 COMMUNICATION JAMMING

Figure A.2 shows the links associated with communication jamming. The jammer is located away from the transmitter and the receiver. Thus, there is a jamming link and a desired signal link. The receiver is designed to receive the desired signal, and if it has a directional antenna, that antenna is oriented toward the desired signal transmitter. Because the jammer is located away from the desired signal transmitter, the target receiver antenna typically has reduced gain in

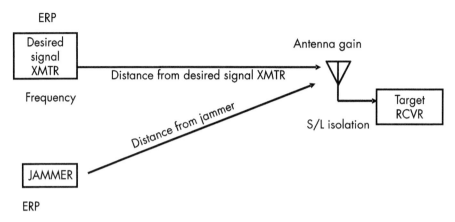

Figure A.2 Communication jamming geometry.

a sidelobe of its antenna pattern in the jammer direction. If the target receiver has a nondirectional antenna, its antenna gain will be the same in the desired signal transmitter and jammer directions.

A.2.1 Successful Communications Jamming

The real test of jammer effectiveness is the thoroughness with which information flow is stopped. This latter factor is commonly tested by having a trained communicator read a text while a trained listener copies down what he or she heard. As the jamming gets worse, the percentage of correct words decreases. One problem with this test is that the human brain allows an operator to pull information from low-quality signals by pattern recognition. We know our language and therefore can complete words or sentences from past experience. One way to overcome this problem is to change key words in the text each time that a message is sent and count only those words for accuracy. Another way is to use random lists of words.

The mechanism by which a communication jammer interferes with communication is injecting an undesired signal into the target receiver along with any desired signals that are being received. The undesired signal must be strong enough that the receiver cannot recover the required information from desired signals. The ratio of the jamming signal (in the receiver) to the desired signal (in the receiver) is called the jamming-to-signal ratio (J/S). This

A.2.2 Communications J/S

The jammer to receiver link and the desired transmitter to receiver link in Figure A.2 can have any of several propagation models. They do not have to have the same propagation model. For this reason, the J/S formulas in this section include general terms for loss. In general, a J/S of 0 dB is adequate for normal noise jamming, but more may be required to overcome EP measures.

For target receivers with nondirectional antenna patterns, in many cases, the receiving antenna of the target receiver has 360° azimuth coverage. Examples of such antennas are whips and monopole antennas. The J/S for targets that have these types of antennas is:

$$J/S = ERP_J - ERP_S - LOSS_J + LOSS_S$$

where J/S is the J/S in decibels, ERP_J is the effective radiated power of the jammer, ERP_S is the effective radiated power of the desired signal transmitter, $LOSS_J$ is the loss from the jammer to the target receiver, and $LOSS_S$ is the loss from the desired signal transmitter to the target receiver.

If a target receiver has a directional antenna, it is good practice to assume that the antenna is oriented toward the transmitter of the signal that it is designed to receive. The direction toward which the antenna is intended to point is called its boresight direction. This is usually, but not always, the direction of maximum gain. The jamming signal is received from some other direction to which the receiving antenna presents a lower antenna gain. In databases of hostile signals, this reduced sidelobe (S/L) gain is stated as: S/L = some number of decibels. This means that the average sidelobe level is this number of decibels below the boresight gain of the antenna. The formula for J/S if the target receiver has a directional antenna is:

$$J/S = ERP_J - ERP_S - L_J + L_S + G_{RJ} - G_R$$

where J/S is the ratio of the jammer power to the desired signal power at the input to the receiver being jammed (in decibels), ERP_J is the effective radiated power of the jammer (in dBm), ERP_S is the effective radiated power of the desired signal transmitter in dBm), L_J is the propagation loss from the jammer to the receiver (in decibels), L_S is the propagation loss from the desired signal transmitter to the receiver (in decibels), G_{RJ} is the receiving antenna gain in the direction of the jammer (in decibels), and G_R is the receiving antenna gain in the direction of the desired signal transmitter (in decibels).

A.2.3 Communications EP

In order to protect communication links from jamming, there are special modulations applied to transmitted signals to spread their modulation. As shown in Figure A.3, this spreading modulation can be removed in an intended receiver, but cannot be removed in a hostile receiver. The spreading modulation significantly reduces the effect of communication jamming and also makes it more difficult for a hostile receiver to detect or receive the spread signal.

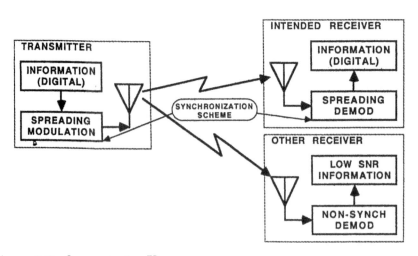

Figure A.3 Communication EP.

A.3 RADAR JAMMING

A radar transmits signals to a target and receives the signals repropagated from that target. These repropagated signals are called skin returns. Radar jamming involves the transmission of signals to the receiver in a radar to interfere with the ability of that receiver to properly receive or process skin return signals. Here, we will divide radar jammers into those that are located on the target platforms being tracked by a radar and those that are remote from that target. If a jammer is on the target vehicle, it is a self-protection jammer; if it is located elsewhere, it is a stand-off jammer.

A.3.1 Successful Radar Jamming

Successful radar jamming is achieved when the ratio of the jamming signal as received by the target radar to the skin return as received by the target radar is adequate to prevent the target radar from acquiring or tracking its target. This ratio is called the J/S.

A.3.2 Self-Protection Jamming

Figure A.4 shows a self-protection jammer. The jammer is located on the target and broadcasts into the boresight of the radar's antenna because the radar has its antenna pointed at the target when the radar is doing its job. The radar antenna may scan away from the target for various reasons, and the pattern of that scan can be a valuable parameter for an intercept receiver to detect. However, when analyzing the J/S of that radar, we consider it only when the radar antenna is on the target.

The radar that is being jammed is called the target radar. The formula for the J/S of a self-protection jammer is

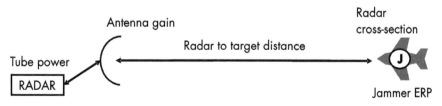

Figure A.4 Self-protection jamming geometry.

$$J/S = ERP_J - ERP_S + 71 + 20 \log R - 10 \log RCS$$

where J/S is the J/S in decibels, ERP_J is the effective radiated power of the jammer in dBm, ERP_S is the effective radiated power of the target radar in dBm, R is the range from the target radar to its target in kilometers, and RCS is the radar cross-section of the target radar's target in m².

A.3.3 Stand-Off Jamming

Figure A.5 shows a stand-off jammer. Note that this jammer is remote from the radar's target. Because stand-off jammers are high value–low inventory assets, they are almost always kept beyond the range of weapons controlled by the target radar being jammed. Because the radar has its antenna boresight aimed at its target and the jammer is remote from that target, the jammer is in a sidelobe of the radar's antenna.

If radar jamming is performed from a satellite, this is clearly stand-off jamming. The formula for the J/S achieved in stand-off jamming is:

$$J/S = ERP_J - ERP_S + 71 + G_S\ G_S + G_M\ G_M - 20 \log R_J + 40 \log R_T - 10 \log RCS$$

where J/S is the J/S in decibels, ERP_J is the effective radiated power of the jammer in dBm, ERP_S is the effective radiated power of the target radar in dBm, G_S is the sidelobe gain of the target radar's antenna in dBi, G_M is the boresight gain of the target radar's main beam in dBi,

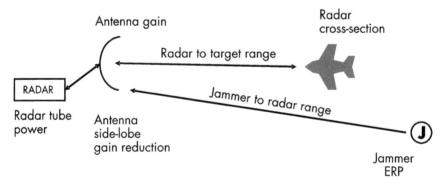

Figure A.5 Remote jamming geometry.

R_J is the range from the target radar to the stand-off jammer in kilometers, R_T is the range from the target radar to its target in kilometers, and RCS is the radar cross-section of the target radar's target in m².

A.3.4 Required J/S

The J/S required for effective radar jamming is a function of the type of jamming employed. If noise jamming is used, 0 dB may be adequate. However, there are also deceptive jamming techniques that can require far greater J/S. Because these deceptive techniques are associated with self-protection jammers and satellites by their nature are stand-off jammers, these techniques are left out of the current discussion.

A.3.5 Radar EP

EP measures in radars are designed to defeat specific types of jamming. In this discussion, we will focus only on a few of these measures that are most appropriate to satellite jamming.

Figure A.6 illustrates the sidelobe suppression technique. It protects the radar from narrow-band sidelobe jamming (particularly frequency modulated noise jamming). There is an auxiliary antenna that has more gain in the sidelobe direction than the sidelobe

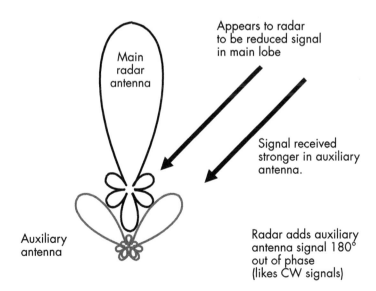

Figure A.6 Radar sidelobe suppression.

gain of the main radar antenna. If the radar detects a stronger signal in the auxiliary antenna than that in the main antenna, the output of the auxiliary antenna has a 180° phase change and is added to the output of the main radar antenna to effectively cancel out the sidelobe jamming signal.

Figure A.7 illustrates sidelobe blanking. It protects the radar from wideband (pulse) jamming in its sidelobes. This technique uses the same type of auxiliary antenna to detect signals that are stronger than in the main radar antenna. When they are present, the whole radar antenna output is blanked during the brief time that the jamming pulse is present.

Figure A.8 illustrates frequency hopping. In this technique, the radar changes the frequency of each pulse in a random pattern. This typically forces a jammer to spread its jamming power over a wide bandwidth to cover all of the transmitted frequencies, thus reducing te J/S to each transmitted pulse.

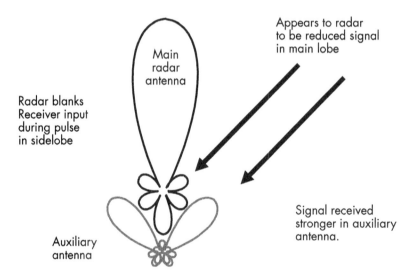

Figure A.7 Radar sidelobe blanking.

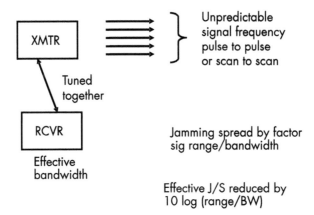

Figure A.8 Frequency-hopping radar.

B

IMPORTANT NUMBERS FOR SPACE EW

- The solar day is 24 hours or 1,440 minutes.
- The sidereal day is 23.9349 hours or 1,436.094 minutes.
- Kepler's third law is $a^3 = C \times P^2$.
  - $C = 36,355,285$ km^3 per min^2.
- Radius of the Earth = 6,371 km.
  - The Earth is proportionally a pretty smooth sphere that can be assumed to be a perfect sphere in most orbital calculations.
- Synchronous satellite period is 23 hours and 56 minutes.
- The 12-hour satellite is 20,241 km high.
- Synchronous altitude is 35,873 km.
- Synchronous satellite semi-major axis is 41,759 km.
- The range from a synchronous satellite to the horizon is 41,348 km.
- The width of the Earth from a synchronous altitude is 17.38°.

Table B.1 shows the parameters of circular orbits with specified periods.

Table B.1
Parameters of Circular Orbits with Specified Periods

p(min)	h(km)	a(km)	rng(km)	dist(km)	p(min)	h(km)	a(km)	rng(km)	dist(km)
90	281	6,652	1,914	1,859	330	9,447	15,818	14,478	7,365
105	1,001	7,372	3,710	3,359	345	9,923	16,294	14,997	7,447
120	1,688	8,059	4,935	4,198	360	10,392	16,763	15,505	7,523
135	2,346	8,717	5,950	4,785	375	10,854	17,225	16,004	7,593
150	2,980	9,351	6,845	5,232	390	11,311	17,682	16,494	7,658
165	3,594	9,965	7,662	5,587	405	11,761	18,132	16,976	7,719
180	4,189	10,560	8,422	5,880	420	12,206	18,577	17,451	7,776
195	4,768	11,139	9,137	6,127	435	12,646	19,017	17,918	7,830
210	5,332	11,703	9,817	6,339	450	13,081	19,452	18,379	7,880
225	5,883	12,254	10,467	6,523	465	13,510	19,881	18,833	7,928
240	6,422	12,793	11,093	6,685	480	13,936	20,307	19,281	7,973
255	6,949	13,320	11,698	6,829	495	14,357	20,728	19,724	8,016
270	7,466	13,837	12,284	6,958	510	14,773	21,144	20,162	8,056
285	7,974	14,345	12,853	7,075	525	15,186	21,557	20,594	8,095
300	8,473	14,844	13,408	7,180	540	15,595	21,966	21,021	8,131
315	8,964	15,335	13,949	7,277	—	—	—	—	—

p (min) is the period of a satellite in minutes. h (km) is the altitude in kilometers for a satellite in a circular orbit. a (km) is the semi-major axis of any satellite in kilometers. *rng* (km) is the range from the SVP to the horizon in kilometers for circular orbit. *dist* (km) is the Earth's surface distance from the SVP to the horizon in kilometers for circular orbit.

C

DECIBEL MATH

C.1 DECIBEL NUMBERS

The math used in this book is mainly "dB math," which is very popular in EW because it allows the convenient comparison of very large numbers such as transmitted signal strength and very small numbers such as received signal strength.

In EW, we spend a lot of time manipulating widely varying signal strength values. We also deal with noninteger powers and roots of numbers. The use of decibel (or dB) forms of numbers and equations greatly simplifies dealing with both of these considerations.

Any number expressed in decibels is logarithmic, which makes it convenient to compare values that may differ by many orders of magnitude. (Note that numbers in nondecibel form are called linear to differentiate them from the logarithmic decibel numbers.) Numbers in decibel form also have the great charm of being easy to manipulate:

- To multiply linear numbers, you add their logarithms.
- To divide linear numbers, you subtract their logarithms.
- To raise a linear number to the nth power, you multiply its logarithm by n.
- To take the nth root of a linear number, you divide its logarithm by n.

To take maximum advantage of this convenience, it is common to put numbers into decibels form as early in the process as possible and to convert them back to linear forms as late as possible (if at all). In many cases, the most commonly used forms of answers remain in decibels, so we can avoid converting back to linear forms altogether.

It is important to understand that any value expressed in decibels units must be a ratio (which has been converted to the logarithmic form). Common examples are:

- Amplifier gain (i.e., the ratio of the output signal strength to the input signal strength);
- Antenna gain (also treated like an amplification ratio, but with some qualifiers);
- Losses (i.e., signal attenuation ratio) when passing through:
  - Cables;
  - Switches (the off position has much more attenuation than the on position, but the on position still has some loss);
  - Power dividers (i.e., ratio of signal power at each output port to input power);
  - Filters.

To create useful equations in decibel form, it is necessary to express absolute values as decibel numbers. Signal strength in units of dBm is the most common example. Since decibel values must always be ratios, a trick is required. The trick is to calculate the ratio of the desired absolute value to some fixed value and then convert that ratio to decibel form. For example, signal strength in dBm is the decibel form of the ratio of that signal strength to 1 mW.

C.2 CONVERSION TO DECIBEL FORM

In the following discussion, we use logarithms, abbreviated "log." In this case, we are using the logarithm to the base 10, which is what you get when you punch the "LOG" key on a calculator.

The basic formula for conversion into decibels is:

$$\text{Ratio (in dB)} = 10 \log(\text{Linear Ratio})$$

For example, 2 (the ratio of 2 to 1) converts to the decibel form as:

$$10 \log(2) = 3 \text{ dB}$$

(it is actually 3.0103 dB, but everyone rounds it to 3 dB) and 1/2 (i.e., 0.5) becomes:

$$10 \log(0.5) = -3 \text{ dB}$$

To make the conversion the easy way (using a calculator), it will have to be a scientific calculator, which has log and 10^x functions.

Another way to talk about converting back to the nonlogarithmic form from the logarithmic form is to use the expression "Antilog" (logarithmic number) in place of $10^{(\text{logarithmic number})}$.

Thus, 3 dB is converted to the nonlogarithmic form as Antilog {3/10}, which is very close to 2.

C.3 ABSOLUTE VALUES IN DECIBEL FORM

As stated above, the most common example of an absolute value expressed in the decibel form is signal strength in dBm. This is the ratio of the signal power to 1 mW converted to the decibel form.

Note that dBm is a particularly important unit because many important formulas in the heart of this book either start or end (or both) with dBm values of signal strength. For example, converting 4W to dBm:

$$4W = 4,000 \text{ mW}$$

$$10 \log(4,000) = 36 \text{ dBm}$$

and

$$1036/10 = 103.6 = 4,000 \text{ (mW)} = 4W$$

C.4 DECIBEL FORMS OF EQUATIONS

Decibel form equations use absolute numbers (usually in dBm) and ratios (in decibels). A typical equation includes only one element in dBm on each side (modified by any number of ratios in decibels), or only ratios in decibels, or differences of two dBm values (which become decibel ratios). One of the simplest decibel form equations is illustrated by the amplifier, which multiplies input signals by a gain factor.

The linear form of the amplifier equation is:

$$P_o = P_I \, G$$

where P_o is the output power, P_I is the input power, and G is the gain of the amplifier.

Both power numbers are in linear units (for example, milliwatts), and G is the gain factor in linear form (for example, 100). If the input power is 1 mW, an amplifier gain of 100 will cause a 100-mW output signal.

By converting the input power to dBm and the gain to decibels, the equation becomes:

$$P_o = P_I + G$$

The output power is now expressed in dBm. Using the same numbers, 1 mW becomes 0 dBm, the gain becomes 20 dB, and the output power is +20 dBm. (This can be converted back to 100 mW in linear units if required.)

This is a very simple case, in which the marginally simpler calculation does not seem worth the trouble to convert to and from decibel forms. Now consider a typical communication theory equation. As shown in Chapter 4, a transmitted signal is reduced by a spreading loss that is proportional to the square of its frequency (F) and the square of the distance (d) that it travels from the transmitting antenna. Thus, the spreading loss is the product of $F2$, $d2$, and a constant (which includes several terms from the derivation). The formula is then:

$$L = K \times F^2 \times d^2$$

In decibel form, F (dB) becomes $10 \log (F)$. F^2 is $2[10 \log (F)]$ or $20 \log (F)$, and d^2 is transformed to $20 \log(d)$ the same way. The constant is also converted to the decibel form, but first it is modified with conversion factors to allow us to input values in the most convenient units and generate an answer in the most convenient units. In this case, K is multiplied by the necessary conversion factors to allow frequency to be input in megahertz and distance to be input in kilometers. When the log of this whole mess is multiplied by 10, it becomes 32.44, which is sometimes rounded to 32. The spreading loss in decibels can then be found directly from the expression:

$$L_s = 32 + 20 \log(F) + 20 \log(d)$$

where L_s is the spreading loss (in decibels), F is the frequency (in megahertz), and d is the distance in kilometers, which most people find much easier to use in practical applications.

It is important to understand the role of the constant in this type of equation. Since it contains unit conversion factors, this equation only works if you input values in the proper units. In this book, the units for each term are given right below the decibel equation every time. You will memorize some of these equations and use them often; be sure you also remember the applicable units.

C.5 QUICK CONVERSIONS TO DECIBEL VALUES

Table C.1 gives some common decibel values with their equivalent linear ratios. For example, multiplying a linear number by a factor of 1.25 is the same as adding 1 dB to the same number in the decibel form (1 mW × 1.25 is the same as 0 dBm + 1 dB, so 1.25 mW = 1 dBm).

This table is extremely useful, because it will allow you to make quick determinations of approximate decibel values without touching a calculator. Here is how it works:

- First, get from 1 to the proper order of magnitude. This is easy, because each time that you multiply the linear value by 10, you just add 10 dB to its decibel value. Likewise, dividing by 10 subtracts 10 dB from the decibel value.

Table C.1
Common Decibel Values

Ratio	Decibel Value	Ratio	Decibel Value
1/10	−10	1.25	+1
1/4	−6	2	+3
1/2	−3	4	+6
1	1	10	+10

- Then use the ratios from Table C.1 to get close to the desired number.

For example, 400 is 10 × 10 × 4. In decibel form, these manipulations are 10 dB + 10 dB + 6 dB. A more common way to look at the manipulation is:

400 is: 20 dB (which gets you to 100) +6 dB (to multiply by 4) = 26 dB and 500 is approximately: 30 dB (= 1,000) − 3 dB (to divide by 2) = 27 dB

Be careful not to be confused by 0 dB. A high-ranking government official once embarrassed himself in a large meeting by announcing that, "The signal is completely gone when the signal to noise ratio gets down to 0 dB." A 0-dB ratio between two numbers just means that they are equal to each other (i.e., have a ratio of 1).

Table C.2 shows the power in dBm for various linear power values. This is a very useful table, and we will use these values many times in examples in other chapters.

Other values that are often expressed in the decibel form are shown in Table C.3.

C.5 QUICK CONVERSIONS TO DECIBEL VALUES

Table C.2
Signal Strength Levels in dBm

dBm	Signal Strength
+90	1 MW
+80	100 kW
+70	10 kW
+60	1 kW
+50	100 W
+40	10 W
+30	1 W
+20	100 mW
+10	10 mW
0	1 mW
−10	100 μW
−20	10 μW
−30	1 μW

Table C.3
Common Decibel Definitions

dBm	= dB value of (Power/1 mW)
dBW	= dB value of (Power/1 W)
dBsm	= dB value of (Area*/1 m^2)*
dBi	= dB value of antenna gain relative to that of an isotropic antenna

*Commonly used for antenna area and radar cross-section.

BIBLIOGRAPHY

Adamy, D. L., *EW 101: A First Course in Electronic Warfare*, Norwood, MA: Artech House, 2001.

Adamy, D. L., *EW 102: A Second Course in Electronic Warfare*, Norwood, MA: Artech House, 2004.

Adamy, D. L., *EW 103: Tactical Battlefield Communication Electronic Warfare*, Norwood, MA: Artech House, 2008.

Adamy, D. L., *EW 104: Electronic Warfare Against a New Generation of Threats*, Norwood, MA: Artech House, 2015.

Baker, R. M. L. Jr., and M. W. Makemson, *Introduction to Astrodynamics*, New York: Academic Press, 1960.

Gibson, J. D., *The Communications Handbook*, Boca Raton, FL: CRC Press, 1997.

Pritchard, W., H. Suyderhoud, and R. A. Nelson, *Satellite Communication Systems Engineering*, 2nd ed., Upper Saddle River, NJ: Prentice-Hall, 1993.

ABOUT THE AUTHOR

Dave Adamy is an internationally recognized expert in electronic warfare (EW), probably mainly because he has been writing the EW-101 columns in the *Journal of Electronic Defense* for many years. He also spent seven years as a systems engineer on reconnaissance satellite payload programs, developing payloads for low Earth and synchronous satellites. He has been an EW professional (proudly calling himself a "Crow") in and out of uniform for over 50 years. As a systems engineer, project leader, program technical director, program manager, and line manager, he has directly participated in EW programs from just above DC to just above light. Those programs have produced systems that were deployed on platforms from submarines to space and met requirements from "quick and dirty" to high reliability.

Mr. Adamy holds BSEE and MSEE degrees, both with communication theory majors, and including several courses on orbit mechanics. In addition to the EW-101 columns, he has published many technical articles in EW, reconnaissance, and related fields, including satellites and applications. He has 17 books in print, including this one. Mr. Adamy teaches EW-related courses all over the world and consults for military agencies and EW companies. He was a long-time member of the National Board of Directors and a past president of the Association of Old Crows.

He has been married to the same long-suffering wife for 59 years (she deserves a medal for putting up with a classical nerd that long) and has four daughters, eight grandchildren, and two great-grandchildren. He claims to be an OK engineer, but one of the world's truly great fly fishermen.

INDEX

A

Absolute values, in decibel form, 215
Airborne hostile jammer, 89
Antenna elevation, 168, 171
Antenna misalignment
  about, 59–60
  loss, 174, 189–90
  transmitting and receiving antennas, 60
Antenna pointing, 166–69
Antenna pointing angles, 169–72
Application link losses, 85
Atmospheric loss
  cause of, 54
  chart, 54–55
  as function of elevation angle, 173
  intercept from radar signal, 154–57
  intercept from synchronous satellite, 182
  LOS loss, 58–59
  through whole atmosphere, 61
  UAV, 191
Azimuth
  calculation, 28
  satellite antenna and ground station, 75–76, 104
  satellite antenna and intercept site, 107
  to target from satellite, 168
  to target hostile transmitter, 171
  transmitting antennas and, 72
  of a vector, 28
Azimuth angle
  calculating, 29–31
  defined, 28, 29
  in direction to ground station, 72
  illustrated, 28

B

Bandwidth
  of command signals, 79
  of data links, 82
  intercept receiver, 110
  jamming and, 208
  satellite payload, 152, 159
  signal, 130, 182
  uplink receiving antenna, 125
Beamwidth, antenna, 108, 125–26, 192

C

Calculations, this book, 4
Circular orbits
  altitude and semi-major axis of, 19, 134
  attitude as function of period, 19
  horizon distances for, 35–36, 136–37
  parameters of, 212
  See also Orbits
Circular polarization, 62
Command links, 78–79
Command spoofing, 92
Communication jamming
  EP, 204
  ERPs, 187
  formulas, 201–4
  Fresnel zone distance formula, 189
  geometry, 187, 202

Communication jamming (continued)
  J/S, 188, 203–4
  link equations, 186–90
  link propagation mode, 189–90
  links and, 188
  network, 186
  successful, 202
Communication signals, 41
The Communications Handbook, 42
Conversion to decibel form, 214–15

D

Data link, 81–82
Data users, links to, 82
Decibel form
  absolute values in, 215
  conversion to, 214–15
  of equations, 216–17
Decibel math, 213–19
Decibel numbers, 213–14
Decibel values
  common, 218, 219
  quick conversions to, 217–18
Directional antennas
  aimed at signal transmitter, 191
  boresight gain, 102
  intercept with, 101–3
Direct sequence spread spectrum, 132
Distance
  from aircraft to ground station, 112
  downlink, 75, 117
  Fresnel zone, 49
  to horizon, calculating, 32–36, 134–36
  to horizon, Earth surface, 141
  to horizon for circular orbits, 35–36
  link, determination, 80
  link, to the horizon, 34–35
  rain loss, 65, 156
  through the rain, 161, 162
  See also Propagation distance
Doppler shift
  calculation in uplink or downlink, 148–49
  formula, 144–45
  general formula for, 148–50
  maximum, general formula for, 148
  receiving site velocity and, 145
  in satellite link, 144–50
  satellite velocity and, 145–48
Downlink distance, 75, 117
Downlink intercept
  about, 96

angle relative to ground station, 103–5
angle relative to hostile intercept site, 105–9
with directional antennas, 101–3
geocentric angle to intercept site, 96–97
intercepted signal quality, 109–10
range to intercept site, 97–99
received signal at intercept site, 109
signal strength at intercept site, 101
Downlinks
  about, 80
  data link, 81–82
  to data users, 82
  Doppler shift calculation in, 148–49
  equation, 80–81
  illustrated, 81
  jamming, 114–19
  jamming links, 82–87
  losses, 82
  telemetry link, 81–82
  See also Satellite links
Duration of target availability
  about, 137–39
  calculating, 137–44
  geocentric viewing angle, 139–41
  impact of movement on Earth, 142
  seeing a point on Earth, 141–42
  viewing time formula, 143–44

E

Earth movement, impact of, 142–43
Earth surface
  area seen by satellite, 140
  distance from satellite SVP to horizon, 160
  distance from SVP to intercept site, 99
  distance to horizon, 141
  location, 21–22
Earth traces
  about, 23
  defined, 23
  of low-Earth satellite, 24
  polar orbit, 25
  satellite, 23–25
  synchronous satellite, 26–27
Eccentricity of orbits, 16
Effective radiated power (ERP)
  defined, 39
  jammers, 122
  jamming of communications network, 187

of radar, 195
uplink, 124
Electronic protection (EP)
  about, 200
  communications, 204
  intercept and, 128
  link jamming and, 129–30
  radar, 207–9
  spoofing and, 128
Elevation
  above local horizon, 76
  antenna, 105, 107, 168, 171
  to Earth satellite, calculation from plane triangle, 65
  of ground station, 108
  from nadir, 139, 153, 185
  satellite from ground station, 76, 116
  to target transmitter, 172
  to threat location, calculating, 31–32
  transmitting antennas and, 72
  of a vector, 28
Elevation angle
  atmospheric loss as function of, 173
  in direction to ground station, 72
  in distance to horizon, 34
  illustrated, 28
  to a threat, 32
Elliptical orbits, 15–17
Error correction codes, 130
EW
  against communications, 37–40
  as dynamic by nature, 42
  equations, inputs into, 4
  formulas from, 199–209
  link vulnerability to, 91–132
  against radars, 37
  signal interception, 28
  space, important numbers, 211–12
EW threats
  azimuth calculation, 29–31
  location of, 28–32
  look angles calculation, 28
  range and elevation calculation, 31–32

F

Flow, this book, 2
Formulas
  communication jamming, 201–4
  intercept, 200–201
  radar jamming, 205
  from SIGINT and EW, 199–209
Frequency hopping, 130, 209

Fresnel zone, 49–50

G

Great circle plane, 140
Ground-based jammers, 93
Ground communications link, jamming, 83–85
Ground radar jamming from space
  asset protection with RCS, 197–98
  duration of jamming, 198
  jammed radar and target, 195–96
  jammer, 196–97
  jamming adequacy, 197
  jamming equation, 197
  jamming geometry, 195
  from satellite, 194
Ground radars, jamming, 85–87
Ground station
  angle relative to, 103–5
  antenna, 102
  distance from aircraft to, 112
  elevation and range to satellite from, 116
  elevation of, 108
  geocentric angle from aircraft to, 113
  propagation distance between receiving satellite and, 125
  range from satellite to, 105
  speed of, 145
  spherical triangle formed by North Pole, SVP, and, 73, 103, 111, 116

H

Horizon
  distances for circular orbits, 35–36
  distance to, calculating, 32–36, 134–36
  Earth surface distance from satellite SVP to, 160
  geometric angle to, 35
  intercept from, 176–80
  plot on Earth, 163–66
  range from satellite to, 160
  range to, 33, 100, 135
Hostile ground-based jammer, 89
Hostile jammers, 115
Hostile links, 87–89
Hostile receiver, 88
Hostile transmitter, 79, 80

I

Intercept
  with directional antenna, 101–3

Intercept (continued)
  downlink, 96–110
  electronic protection and, 128
  illustrated, 92
  successful, 91–92
Intercepted downlink signal, quality of, 109–10
Intercept formulas
  intercept link equation, 201
  received signal quality, 201
  successful intercept, 200–201
Intercept from radar signal
  about, 151–52
  atmospheric and rain loss, 154–57
  duration to see signal and, 162–63
  link loss, 152
  link margin, 159
  from low-Earth satellite, 151–63
  payload receiving signal and, 157–59
  receiver sensitivity, 159
  receiving signal from its horizon and, 159–62
Intercept from synchronous satellite
  about, 180
  with satellite directly overhead, 182
  with satellite on the horizon, 180–82
Intercept from the horizon
  about, 176
  duration of, 179–80
  LOS loss, 176–77
  rain distance, 177–78
  total link loss, 179
Intercepting uplinks, 110, 111
Intercept links, 79–80, 94
Intercept of Earth surface target
  about, 166
  antenna pointing, 166–69
  antenna pointing angles, 169–72
  intercept from horizon, 176–80
  intercept link equation, 169–72
  link losses, 172–76
  with narrow-beam receiving antenna, 166–80
  rain loss, 175
  receiver sensitivity, 176
Intercept receiver
  bandwidth, 110
  propagation distance between satellite and, 98
  received power at, 113–14
Intercept site
  downlink signal strength at, 101

Earth surface distance from SVP to, 99
geometric angle from satellite to, 96–97
hostile, angles relative to, 105–9
range from satellite to, 97–99
received signal at, 109
satellite above horizon and, 99–101

J

Jammer link, 118
Jammers
  ERP, 122
  ground-based, 93
  ground radar jamming from space, 196–97
  hostile, 115
  LOS loss, 118
  offset, 126–27
  operating against satellite downlink, 93
Jammer-to-signal ratio (J/S)
  for communication jamming, 95, 188
  communications, 203–4
  defined, 83–84, 94
  formula, 84–85
  jamming downlinks, 118–19
  jamming satellite uplinks, 127–28
  radar jamming, 207
  satellite communication jammer, 85
  satellite radar jammer, 86
Jamming
  bandwidth and, 208
  of communications network, 190–93
  duration of, 198
  of ground radar from space, 194–98
  link, electronic protection and, 129–30
  protection against, 130–32
  from satellite, 183–86
  satellite vulnerability and, 94–95
  uplink, 93–94
Jamming antenna, 122
Jamming downlinks
  about, 114–15
  jammer link, 118
  J/S formula, 118–19
  satellite downlink, 115–18
Jamming from space
  of communications network, 186–90
  of ground radar from space, 194–98
  microwave digital data link, 190–93
  from satellite, 183–86
Jamming links
  about, 82–83
  ground communications, 83–85

ground radars, 85–87
satellite uplinks, 120–22
Jamming satellite uplinks
  about, 119
  antenna beamwidth, 125–26
  jammer ERP, 122
  jammer offset, 126–27
  jamming antenna gain, 122
  jamming link, 120–22
  jamming link loss, 122
  J/S formula, 127–28
  satellite uplink, 123–24
  uplink ERP, 124
  uplink loss, 125
  uplink receiving antenna gain and bandwidth, 125
  uplink transmitter antenna gain, 124

K

Kepler, Johannes, 15
Keplerian ephemeris, 15
Kepler's third law, 18, 26, 163, 179, 211
Knife-edge diffraction (KED)
  about, 42
  attenuation, 51, 54
  estimation, 51
  factor, 52
  geometry, 52
  graphic determination, 53
  LOS path and, 53

L

Linear polarization, 60, 62
Link geometry
  ground station location in, 70–71
  looking down, 72–74
  looking up, 75–76
  orbit numbers and, 74–75
  SVP in, 69, 70
  See also Satellite links
Link jamming, electronic protection and, 129–30
Link losses
  application, 85
  intercept from radar signal, 152–54
  intercept of Earth surface target, 172–76
  jamming, 83, 122
  one-way link, 40
  satellite to target, 174–75
  space-related, 94

Link losses intercept of Earth surface target, 172–76
Link margin, 159
Link vulnerability, 91–132
Look angles, 5, 28–29, 32, 103, 105, 133, 166
LOS loss
  about, 57–58
  atmospheric loss, 58–59
  intercept from radar signal, 154
  intercept from the horizon, 176–77
  jammer, 118
LOS propagation
  about, 42–44
  alternate forms of equation, 45
  equation, 44–45
  illustrated, 43

M

Microwave digital data link, jamming, 190–93
Misalignment, antenna, 59–60, 174, 189–90

N

Napier's rules, 10–11, 108
National Reconnaissance Office (NRO), 4
Numbers
  decibel, 213–14
  for space EW, 211–12

O

One-way link
  elements of, 38–39
  equation, 38
  EW against communications and, 37–40
  illustrated, 38
  link loss, 40
  output to receiving antenna, 39–40
Orbits
  circular, 19, 35–36, 134, 136–37, 212
  eccentricity of, 16
  elliptical, 15–17
  great circle plane, 140
  low, 1
  mechanics of, 15–36
  polar, 25
  size and period relationship, 18–20
  synchronous, 1
  trade-off in selection of, 2

P

Payload functions, 78
Payload receiver, 79–80
Period, orbital, 19, 141, 212
Plane triangle
  elevation from Earth satellite from, 65
  examples in problems, 11–13
  illustrated, 6
  law of cosines for, 31–32, 172
  law of cosines for angles, 7
  law of cosines for sides, 6, 73, 107
  law of sines, 6, 73, 117
  right, 7, 163
  sum of three internal angles, 118
Plane trigonometry, 5–7
Polarization
  circular, 62
  linear, 60, 62
  rotation of, 158–59
  satellite link, 63
  small antennas and, 62
Polarization loss, 60–62, 158
Polar orbit, 25
Propagation
  introduction to, 37
  LOS, 42–45
  one-way link, 37–40
  two-ray, 46–49
Propagation distance
  about, 22–23
  between receiving satellite and ground station, 125
  between receiving satellite and jamming transmitter, 121
  between satellite and intercept receiver, 98
  from satellite to target, 187
Propagation in space
  about, 57
  LOS loss, 57–60
  polarization loss, 60–62
  rain loss, 62–67
Propagation loss models
  about, 40–42
  atmospheric loss, 54–55
  complex reflection environment, 50
  Fresnel zone and, 49–50
  KED, 50–54
  LOS propagation, 42–45
  rain and fog attenuation, 55–56
  selection of, 42
  two-ray propagation, 46–49

very low antennas and, 49
Protection against jamming, 130–32

R

Radar cross section (RCS), 197–98
Radar jamming
  about, 205
  EP, 207–9
  required J/S, 207
  from satellite, 194
  self-protection, 205–6
  stand-off, 206–7
  successful, 205
Radar sidelobe blanking, 208
Radio propagation. *See* Propagation
Rain distance, 177, 178
Rain loss
  about, 62–64
  distance, 65, 156
  as function of frequency, 63
  heavy rain, 157, 175
  intercept from radar signal, 154–57
  intercept from synchronous satellite, 182
  intercept of Earth surface target, 175
  UAV, 191
Range
  to horizon, 33, 135
  to horizon from satellite, 100
  to point on Earth, 139
  propagation, intercept link, 113
  rate of change of, 148
  to satellite from ground station, 116
  from satellite to ground station, 105
  from satellite to horizon, 160, 164
  from satellite to intercept site, 97–99
  to synchronous satellite, 181
  to threat, 153
  to threat from satellite, 185
  to threat location, calculating, 31–32
Receiver sensitivity
  intercept from radar signal, 159
  intercept of Earth surface target, 176
Receiving site velocity, 145
Reflection, complex environment, 50, 51
Remote jamming geometry, 206–7
Right plane triangle, 7, 163
Right spherical triangle, 10, 109, 126, 149–50

S

Satellite downlink, 115–18

INDEX

Satellite Earth trace
  about, 23–24
  as path of SVP, 24
  polar view of, 24
  synchronous, 26–27
Satellite ephemeris
  elements of, 16–17
  location definition, 17
  table, 16
Satellite links
  attacks on, 128–30
  command links, 78–79
  data link, 81–82
  to data users, 82
  Doppler shift in, 144–50
  downlinks, 80–87
  electronic protection of, 128–32
  hostile links, 87–89
  intercept links, 79–80
  jamming links, 82–87
  link geometry, 69–76
  polarization, 63
  protection against jamming, 130–32
  signal carried by, 78
  telemetry link, 81
  uplinks, 76–80
  uplinks, jamming, 119–28
  vulnerability, 91–132
Satellite payload, 77, 79–80, 81, 151–52, 157–59, 184, 193
Satellites
  communication and, 1
  disadvantages of, 1
  duration of target availability from, calculating, 137–44
  Earth trace of, 23–25
  jamming from, 183–86
  location relative to the Earth, 15
  radar jamming from, 194
  velocity, 145–48
Self-protection jamming, 205–6
Sidelobe blanking, 208
Signal intelligence (SIGINT)
  formulas from, 199–209
  satellite programs, 3–4
Space-related link losses, 94
Spectrum spreading, 130, 132
Spherical triangle
  defined, 7
  examples in problems, 11–13
  formation illustration, 21
  illustrated, 8

law of cosines for angles, 9–10
law of cosines for sides, 9, 74–75, 104, 106, 112, 165
law of sines, 9, 30, 166, 170
Napier's rules and, 10–11
North Pole, ground site, and ascending node, 147
North Pole, satellite SVP, and ground station location, 123
North Pole, satellite SVP, and transmitter location, 120
North Pole, SVP, and ground station, 73, 103, 116
North Pole, SVP, and intercept site, 106
North Pole, SVP, and location on Earth, 138
North Pole, SVP, and receiver location, 21
North Pole, SVP, and target at longitude and latitude, 167, 170
North Pole, SVP, and threat location, 29, 152, 184
North Pole, SVP of the aircraft, and ground station, 111
right, 10, 109, 126, 149–50
sides of, 8
trigonometric relationships in, 8–11
Spherical triangle for sides, 30
Spherical trigonometry, 5–13
Spoofing
  command, 92
  electronic protection and, 128
  illustrated, 92
  link, 94
  successful, 92
Spreading loss, 160, 172–73
Stand-off jamming, 206
Sub-vehicle point (SVP)
  defined, 11–12, 20
  Earth trace of satellite as path of, 25
  fixed, 26
  illustrated, 20
  in link geometry, 69, 70
  in spherical triangle, 12
  synchronous satellite, 26
Synchronous satellite
  Earth trace, 26–27
  image of, 27
  intercept from, 180–82
  orbit, 1
  range to, 181
  SVP, 26

T

Telemetry link, 81
Trigonometry
  plane, 5–7
  spherical, 5–13
Two-ray propagation
  decibel formula, 46–47
  dominant loss effect, 46
  loss, determination of, 47
  loss variation, 46
  minimum antenna height for, 48–49

U

Unmanned aerial vehicle (UAV) scenario
  about, 190
  antenna beamwidth, 192
  antenna gain, 192
  atmospheric and rain losses, 191
  illustrated, 191
  link gain, 191–92
Uplink jamming, 93–94
Uplink receiving antenna gain and bandwidth, 125

Uplinks
  command links, 78–79
  Doppler shift calculation in, 148–49
  emitters, 76–77
  equation, 77
  ERP, 124
  geometry, 77
  intercepting, 110–14
  intercept links, 79–80
  link distance, 80
  losses, 77, 125
  satellite, jamming, 119–28
  See also Satellite links
Uplink transmitter antenna gain, 124

V

Viewing time formula, 143–44

The Artech House Electronic Warfare Library

Dr. Joseph R. Guerci, Series Editor

Activity-Based Intelligence: Principles and Applications, Patrick Biltgen and Stephen Ryan

Advances in Statistical Multisource-Multitarget Information Fusion, Ronald P. S. Mahler

Antenna Systems and Electronic Warfare Applications, Richard A. Poisel

Electronic Intelligence: The Analysis of Radar Signals, Second Edition, Richard G. Wiley

Electronic Warfare for the Digitized Battlefield, Michael R. Frater and Michael Ryan

Electronic Warfare in the Information Age, D. Curtis Schleher

Electronic Warfare Receivers and Receiving Systems, Richard A. Poisel

Electronic Warfare Signal Processing, James Genova

Electronic Warfare Target Location Methods, Richard A. Poisel

Emitter Detection and Geolocation, for Electronic Warfare, Nicholas A. O'Donoughue

EW 101: A First Course in Electronic Warfare, David L. Adamy

EW 103: Tactical Battlefield Communications Electronic Warfare, David L. Adamy

EW 104: EW Against a New Generation of Threats, David L. Adamy

EW 105: Space Electronic Warfare, David L. Adamy

Foundations of Communications Electronic Warfare, Richard A. Poisel

High-Level Data Fusion, Subrata Das

Human-Centered Information Fusion, David L. Hall and John M. Jordan

Information Warfare and Organizational Decision-Making, Alexander Kott, editor

Information Warfare and Electronic Warfare Systems, Richard A. Poisel

Information Warfare Principles and Operations, Edward Waltz

Introduction to Communication Electronic Warfare Systems, Richard A. Poisel

Introduction to Electronic Defense Systems, 3rd Edition, Filippo Neri

Introduction to Modern EW Systems, 2nd Edition, Andrea De Martino

Knowledge Management in the Intelligence Enterprise, Edward Waltz

Mathematical Techniques in Multisensor Data Fusion, Second Edition, David L. Hall and Sonya A. H. McMullen

Military Communications in the Future Battlefield, Marko Suojanen

Modern Communications Jamming Principles and Techniques, Richard A. Poisel

Practical ESM Analysis, Sue Robertson

Principles of Data Fusion Automation, Richard T. Antony

RF Electronics for Electronic Warfare, Richard A. Poisel

Stratagem: Deception and Surprise in War, Barton Whaley

Statistical Multisource-Multitarget Information Fusion,
   Ronald P. S. Mahler

Tactical Communications for the Digitized Battlefield, Michael Ryan
   and Michael R. Frater

Target Acquisition in Communication Electronic Warfare Systems,
   Richard A. Poisel

For further information on these and other Artech House titles, including previously considered out-of-print books now available through our In-Print-Forever® (IPF®) program, contact:

Artech House	Artech House
685 Canton Street	16 Sussex Street
Norwood, MA 02062	London SW1V 4RW UK
Phone: 781-769-9750	Phone: +44 (0)20-7596-8750
Fax: 781-769-6334	Fax: +44 (0)20-7630-0166
e-mail: artech@artechhouse.com	e-mail: artech-uk@artechhouse.com

Find us on the World Wide Web at: www.artechhouse.com